PROJECT PRE-CHECK
FastPath

The Project Manager's Guide to Stakeholder Management

R. Andrew Davison

Order this book online at www.trafford.com
or email orders@trafford.com

Most Trafford titles are also available at major online book retailers.

Printed in the United States of America.

ISBN: 978-1-4669-1873-3 (sc)
ISBN: 978-1-4669-1874-0 (e)

Trafford rev. 04/19/2012

 www.trafford.com

North America & international
toll-free: 1 888 232 4444 (USA & Canada)
phone: 250 383 6864 ♦ fax: 812 355 4082

Cover design and graphics by Egad Design

TABLE OF CONTENTS

FIGURES

TABLES

In Recognition

To all project managers who work tirelessly to build and watch over the capability of the stakeholder group

PART I
Introduction

- o Why PM's need to read this book
- o How this book is organized
- o Acknowledgements
- o I'd like to hear from you

"It is amazing what you can accomplish if you do not care who gets the credit"

Harry S. Truman (1884-1972)

Is it possible for a project to prosper and be acknowledged as a resounding success if it exceeds initial estimates, takes longer to deliver than first planned, and delivers less functionality than originally hoped for? Absolutely! Can the stakeholders in such a venture enjoy their just rewards? Of course! Is it possible for a project to miss all of its initial targets and still be considered a triumph? Without a doubt! The answer to this paradox lies in the decision-making process and who gets to call the shots.

Change is amorphous. The exact end result is a dream. The way to that dream is a cauldron of competing interests, perceptions, personalities and priorities. A change doesn't come out of someone's head fully formed. It doesn't emerge from a planning meeting signed, sealed and delivered. It has to be molded, shaped, poked and prodded into its ultimate form by the people who have to live with the results—the stakeholders. What's a stakeholder? In the context of Project Pre-Check, it is a decision maker who:

- o Directs an organization that initiates, is affected by or is charged with managing all or part of a change,
- o Has the authority and responsibility to set direction, establish priorities, make decisions, commit money and resources and
- o Is accountable for delivering the planned benefits on budget and on plan.

Stakeholder involvement and commitment is one of the most important ingredients for successful business and technology change. Without it, a project is doomed. The Standish Group has been looking at the factors which most contribute to project success since 1994. In their CHAOS studies, they have found two factors always at the top:

- o **User Involvement!**
- o **Executive Management Support!**

User Involvement and *Executive Management Support* have alternated as number 1 and number 2 from 1994 through 2011. There is no doubt what is needed to ensure a successful change! Interestingly, *Clear Vision and Objectives* has been in the number three spot for most of those years. And, where do clear vision and objectives come from? *User Involvement* and *Executive Management Support*!

Introduction

That's why Project Pre-Check and Project Pre-Check *FastPath* are based on three building blocks fundamental to project success: active and ongoing stakeholder involvement and agreement, a Decision Framework that helps stakeholders form and articulate a common vision for the planned change, and processes that map the way from inception through completion.

Professor Ryan Nelson of the University of Virginia has been studying project performance results over a number of years. In an article published in the June 2007 issue of the MIS Quarterly Executive, he identified the top ten reasons for failure garnered from 99 project reviews. He also identified well established best practices that would have helped the projects avoid their dismal fates.[1] His findings are summarized in the table below. The Project Pre-Check *FastPath* column is mine.

Classic Mistakes	Best Practices										Project Pre-Check FastPath
	Agile Development	Communication Plan	Estimate-Convergence Graph	Joint Application Development (JAD)	Comprehensive Project Charter	Project Management Office	Retrospectives	Staged Delivery	Stakeholder Assessment	Work Breakdown Structure	
Poor estimation and/or scheduling	X		X		X	X	X	X		X	X
Ineffective stakeholder management		X		X	X	X			X		X
Insufficient risk management			X		X	X	X	X			X
Insufficient planning			X		X	X	X			X	X
Shortchanged quality assurance	X			X				X			X
Weak personnel and/or team issues	X	X				X	X	X			X
Insufficient project sponsorship		X		X	X	X			X		X
Poor requirements determination	X			X						X	X
Inattention to politics		X			X	X			X		X
Lack of User involvement	X	X		X					X	X	X

Table 1—Classic Mistakes and Best Practices Matrix

Notice that *Ineffective stakeholder management* is number 2 on the list, *Insufficient project sponsorship* is number 7 and *Lack of User involvement* is number 10. In terms of the best practices that can be leveraged to help avoid the classic mistakes, *FastPath* addresses them all, and a wealth of others besides. No wonder! Project Pre-Check is designed specifically to help stakeholders leverage best practices to deliver major change successfully.

Let's look an example that illustrate the power of engaged stakeholders. For those of you who have had a house built from scratch, and for those that haven't, you know there are thousands of decisions that need to be made, from the fundamental ones dealing with location, lot size and shape and the size and structure of the buildings to selection of hardware for the kitchen cabinets, lighting fixtures and door hardware and everything in between. Every one of these decisions can influence the final cost by thousands of dollars, up or down. The examples below are real life situations. Just substitute your project of choice for the house building exercise to relate them to your own major change experiences.

Couple A, who I'll call the Dissatisfieds, went through the building process and was very unhappy with the end result. Couple B, the Satisfieds, went through the building process and was thrilled with the results. In both cases, the final cost was well over budget but work was completed on target. The Dissatisfieds received everything they wanted in terms of the specified form and features. The Satisfieds actually ended up with a somewhat smaller house than originally planned. Why the difference in attitude towards the final outcome? Stakeholder involvement in the decision-making process from the beginning to the end made all the difference!

There were seven stakeholders involved in both projects: the homeowners, husband and wife (sponsors), the architect and contractor (change agents) and the neighbours on either side plus the building inspector (targets). In the Dissatisieds' situation, most of the decisions were made by the architect and implemented by the contractor. They involved the owners minimally and the neighbours not at all. The owners didn't ask, the architect and contractor didn't tell! When the over budget bills started coming in, the owners directed their wrath at the architect who blamed it on the contractor. The neighbours directed their wrath at the owners for the constant noise, dirt and dust and the location and size of the buildings. One of the neighbours actually sold their house and moved away. The building inspector lost most of his hair and spent many sleepless nights trying to deal with and resolve issues after the fact. Relationships were frayed all around. It was not a successful result!

In the Satisfieds' situation, the owners were actively involved with the other stakeholders from the planning stages on. When there was an issue that could affect the cost, environment or delivery date, they were all engaged and a collective decision was made. Yes, they spent more than planned and delivered less than planned. But, they could point to the incremental costs and smaller house with confidence that they had, collectively, made hundreds of right decisions along the way leading to the end result that they wanted. The builder was pleased with the process and the outcome and used the Satisfieds as a reference for prospective clients. The neighbours were involved throughout and happy with the result. The building inspector was a constructive and contented partner. And of course, the owners were ecstatic! Active stakeholder involvement! It made all the difference.

How do you go about creating that winning stakeholder group? There are thousands of organizations that focus on effective project and change management, lots of books and a veritable avalanche of periodicals and articles about both subjects. But there are few, if any, tools that focus exclusively on stakeholders!

The original Project Pre-Check provided a unique practice for stakeholders to build an effective stakeholder group. It was easy to learn, easy to apply and quick to deliver value. Stakeholders could garner meaningful insight into a project's strengths and weaknesses in a matter of days. The results provided a comprehensive platform for both pro-active and remedial collective action that ensured the focus was always on the real goal—delivering targeted benefits to the organization.

Project Pre-Check *FastPath* builds on that legacy but focuses more on change agents or project managers, a stakeholder role, and one of the key positions charged with navigating the treacherous and turbulent waters of project delivery. Project Pre-Check *FastPath* provides a streamlined, five step process that can be used effectively on new projects and in progress initiatives, a compressed Decision Framework covering the 50 most frequently relevant Decision Areas and real life case studies to illustrate how best to apply Project Pre-Check to deliver optimum value.

I refer to both Project Pre-Check and *FastPath* in this book. Project Pre-Check is generally used in reference to the overall principles and practices. Project Pre-Check *FastPath*, or just *FastPath*, is used to refer to features and characteristics unique to the practice.

"Those who do not remember the past are condemned to repeat it"

From The Life of Reason by George Santayana

Project Pre-Check *FastPath* supplies history's wisdom to project managers managing major business and technology changes.

Why PM's Need to Read This Book

If you are a project manager, or change agent, you are a key decision maker and thus a project stakeholder. Also, you may be the only full time stakeholder on the project. All the other stakeholders undoubtedly have day jobs in addition to their project responsibilities. In fact, that's why they are sponsors, targets and champions. If they didn't have responsibility for the organizations, people, processes and functions involved with and affected by the planned change, they wouldn't be stakeholders.

Along with the other stakeholders, as a PM, you are responsible for making sure that the project progresses smoothly to a successful conclusion. That entails developing, guiding and manipulating a myriad of factors including:

- o   The project plan
- o   Time and cost estimates
- o   Skill numbers and needs
- o   Resource acquisition and assignment
- o   Quality plans
- o   Risk plans
- o   Change requests
- o   Project issues
- o   Progress tracking and reporting
- o   Etc., etc., etc.

You know the challenge! In fact, the original Project Pre-Check Decision Framework includes 125 Decision Areas as a starting point for guiding stakeholder decision-making on major business and technology changes.

The interesting thing about project dynamics though, is that you as a PM can only decide HOW to deliver. Decisions relating to what the change or project is trying to achieve and what's impacted as a result, when it needs to be delivered, who is affected, why it's needed and

Introduction

where it will be implemented are usually in the sponsor's hands, with assistance from the other stakeholders.

So, you as a PM need a heap of help from the other stakeholders to be successful. In fact, as mentioned, stakeholder engagement and user participation are at the top in the vast majority of top ten lists of factors that contribute to project success.

I'll go even further and suggest that projects that screw up one way or another (over budget, late, incomplete functionality and/or poor quality) do so because their stakeholder groups were ineffective, not because the PM blew it or because of a technology problem or whatever other excuses are offered for project failure. Where I do hold the PM accountable is in facilitating the formation and operation of the stakeholder group. In my view, that is the PM's number one challenge!

Yet, many project managers are reluctant to demand the necessary time and attention from the senior executives who usually fill the sponsor and target roles. However, that's exactly what has to happen! For projects to be successful, sponsors, targets and champions, along with the change agents, need to allocate appropriate amounts of their time, attention and effort. That's what Project Pre-Check *FastPath* is designed to do—to help project managers get stakeholders involved, engaged and collaborating from project inception to the completion celebration.

Project Pre-Check *FastPath* gives project managers the tools to:

o   Ensure the active involvement and agreement of stakeholders.
o   Leverage industry and organizational best practices quickly and effectively.
o   Evolve to take advantage of new industry or organization learnings that can improve future project performance.
o   Define and implement the change to maximize the value delivered.

How This Book is Organized

This book is organized to help project managers understand Project Pre-Check *FastPath*, apply the practice to their projects and adapt it to the unique needs that are almost always a part of project life.

There are a number of case studies throughout the book to help you apply the *FastPath* practices. The case studies have a shaded background to help you identify them quickly.

Part I—Introduction, describes the rationale for the book, its organization, the players that will benefit from the material presented and the events and people that shaped the ideas and content.

Part II—Overview, describes the context for Project Pre-Check *FastPath*. It includes opinions on what makes projects successful, how *FastPath* can help, an introduction to the Project Pre-Check building blocks and how they can be used to ensure project success.

Part III—Stakeholders, offers guidance on forming and managing the stakeholder group. It also provides tips for coping with stakeholder changes and what to do if you join an in-flight project.

Part IV—the Project Pre-Check *FastPath* Process, provides step by step coverage for each of the five steps—Identify Stakeholders, Engage Stakeholders, Assess Decision Areas, Monitor Agreement and Guide Completion.

Part V—Decision Framework, includes a subset of the Project Pre-Check Decision Area catalogue which describes the Domains, Factors and Decision Areas that stakeholders need to consider when assessing, planning and controlling a change.

Introduction

Appendices—include listings of a variety of industry sources and best practices that provide the foundation for this book.

Online—Excel templates of the forms and questionnaires referenced in this book are available for download at www.projectprecheck.com. Also, use the site for information or assistance or to contribute your experiences, findings and suggestions.

Acknowledgements

Project Pre-Check *FastPath* is the product of numerous suggestions and contributions from project managers and others who have reviewed and used Project Pre-Check in its original form. Thanks to all who participated. Thanks also to all those fearless change agents who contributed case studies and examples that serve to demonstrate the value of Project Pre-Check to manage change effectively.

I'd Like to Hear from You!

Project Pre-Check *FastPath* is a work in progress. It's a framework that you will adapt to each unique circumstance to achieve your project goals. Your experiences and insights are needed to help others, to improve the processes and keep the best practices evergreen. Tell me what you think, what worked, what didn't, what should be added, deleted or changed at www.projectprecheck.com. There, you'll also be able to download the Project Pre-Check and *FastPath* templates and check out the latest posts on case studies, best practices and stakeholder management. Thanks and great success!

R. A. (Drew) Davison
Kirkfield, Ontario
February 2012

PART II
Overview

- Why Projects Succeed
- How *FastPath* Helps
- Project Pre-Check Building Blocks
- Using Project Pre-Check *FastPath*

Agnes Allen's Law: Almost anything is easier to get into than out of.

Implementing major business and technology change successfully is a formidable undertaking. Examples of ventures cancelled outright, of projects that failed to deliver to expectations, of blown budgets, of questionable quality, of missed deadlines are trumpeted in the press daily.

- o According to the Standish Group, over two thirds of projects fail or are seriously challenged.[2]
- o The U.S. Government's Office of Management and Budget, the federal government agency that evaluates the effectiveness of federal programs, policies, and procedures, found in a March 2003 study, that "771 projects included in the fiscal 2004 budget—with a total cost of $20.9 billion—are currently at risk."[3]
- o Morgan Stanley reported that between 2000 and 2002, U.S. companies spent $130 billion on software and hardware that they ultimately didn't need to support their businesses.[4] According to figures by Gartner, this number jumps to $540 billion on a global scale.[5]

Unfortunately, there are thousands of past and current project failures costing the organizations involved unconscionable amounts in wasted time, money and opportunity. And, unfortunately, there will continue to be thousands of major business and technology change initiatives that will fail in the future.

Why Projects Succeed

Projects still fail because it is difficult and time consuming to try and apply general rules and principles to specific changes. Projects still fail because the stakeholders involved in a change don't actually know why their projects failed or will fail. Projects still fail because the stakeholders involved don't specifically know what should be done to prevent their particular project from failing. And, projects still fail because most stakeholders don't really understand the role they need to play to ensure success!

Google "why projects fail" on any given day and you'll probably get over 15 million hits trying to answer the question. Do the same thing with "why projects succeed" and you'll probably get fewer but still significant results. Scan the available literature and you'll find thousands of learned discourses on the subject with diverse opinions on the causes of project failure and how to achieve success. For example, here's a small sample on how to avoid failure and/or how to ensure your project succeeds:

- o   In his book Death March: Managing "Mission Impossible" Projects, Edward Yourdon states "the solution will involve issues of peopleware, processes and methodologies, as well as tools and technologies."[6]
- o   Jennifer Stapleton, in her book about DSDM, the dynamic systems development method, states "DSDM is about people, not tools. It is about truly understanding the needs of the business and delivering solutions that work—and delivering them as quickly and as cheaply as possible"[7]
- o   In the Project Management Institute's Guide to the Project Management Body of Knowledge, the authors write that "the knowledge and practices described are applicable to most projects most of the time, and that there is general agreement that the correct application of these skills, tools, and techniques can enhance the chances of success over a wide range of different projects. Good practice does not mean that the knowledge described should always be applied uniformly on all projects."[8]

o The I&IT Task Force was commissioned by the Government of Ontario, Canada in 2004 to determine "why some very large I&IT projects struggle, and sometimes fail and to recommend corrective actions." The report, issued in July, 2005, found that "Large IT projects rarely fail due to IT problems alone. In fact, most projects that struggle are engaged in major business transformation, while smaller, more routine IT projects are more likely to complete their project life-cycle without issue. Major business transformation in the Ontario government is often treated merely as an IT initiative, as opposed to the complex organizational change management challenge that it actually is."[9] The report included sixteen major recommendations for improving project success rates.

o The Office of Government Commerce (OGC) in the UK cite the following causes for project failure:

- Design and definition failures where the scope of the programme and / or project(s) are not clearly defined and required outcomes and /or outputs are not described with sufficient clarity.
- Decision making failures due to inadequate level of sponsorship and commitment to the programme and / or project(s), i.e. there is no person in authority able to resolve issues.
- Programme and Project discipline failures, including weak arrangements for managing risks and inability to manage change in requirements.
- Supplier management failures, including lack of understanding of supplier commercial imperatives, poor contractual set-up and management.
- People failure, including disconnect between the programme and / or project(s) and stakeholders, lack of ownership, cultural issues.[10]

There's a huge body of opinion out there about how to implement change successfully, how to avoid project failure and how to achieve maximum benefits at minimum cost and risk. Who's right? Well, actually, all of them are right some of the time. The difficulty is—where do you start? What can a PM do:

o To help a business or IT executive ensure the successful implementation and operation of a new initiative in a tight timeframe, with the quality the clients demand, at a cost the organization can afford and a return that provides the expected value?

o To help customers ensure their operations will continue to survive and thrive when they discover an important supplier is undertaking a major new change initiative that will affect the way they operate?

Overview

- o To help a business or IT executive respond when informed that one of his or her peers is launching a major project that will dramatically change the way their organization operates?
- o To help project staff deliver on the invariably conflicting demands placed on them by other stakeholders?
- o To assist consultants and contractors when they find the project they're on lacks direction, stakeholder commitment and agreement?
- o To help vendors ensure that the products and services they sell to clients deliver the expected value on time and budget?

The answer? Use Project Pre-Check *FastPath*. It gives you the tools to form, engage and support the ongoing operation of a committed, effective stakeholder group. And, it also gives you guidance when the stakeholder group isn't working as well as you'd like it to work.

How FastPath Helps

"A journey of a thousand miles must begin with a single step."

Chinese proverb

A scan of the business and technology press yields numerous opinions on how to implement change successfully. Each opinion usually reflects a particular point of view—from business school academics to methodology and technology vendors to project management practitioners. However, there are a number of success factors that are common on most lists. The most frequently referenced factors are summarized by the Standish Group in their Chaos Ten, in order of the greatest influence on a project's success:[11]

- o Executive Support
- o User Involvement
- o Experienced Project Manager
- o Clear Business Objectives
- o Minimized Scope
- o Standard Software Infrastructure
- o Firm Basic Requirements
- o Formal Methodology
- o Reliable Estimates
- o Other

According to the Chaos study findings, "although no project requires all ten factors to be successful, the more factors that are present in the project strategy, the higher the level of confidence."[12]

All of the lists of conditions that pundits tell us must be met to deliver major change successfully tend to omit or gloss over one critical factor: the need for stakeholders to make lots of tough, informed choices throughout the course of the change. Project Pre-Check

Overview

FastPath ensures that the conditions necessary for success are in place. It achieves that though the following mechanisms, or building blocks:

o   Stakeholders—to ensure the key decision makers are actively involved and in agreement throughout the change.
o   Processes—to ensure that the roadmap to success is applied consistently on each change.
o   Decision Framework—to ensure that a broad perspective is brought to bear on the factors that can contribute to project success and the factors that may be affected by the planned change so that stakeholders can make specific decisions on the changes required to achieve the desired results.

The Project Pre-Check *FastPath* process is the recipe, the Decision Areas are the ingredients! The stakeholders are the chefs!

Project Pre-Check is based on this premise:

If the stakeholders for a given change are actively involved in and agree with each decision, and all the vital decisions are addressed, the project will be successful.

FastPath covers stakeholder involvement from the inception of the change process through to the realization of planned benefits. It addresses controls and practices that stakeholders can leverage to form and guide the change to the intended conclusion or to change direction or even cancel the initiative if conditions warrant in time to avoid undue damage.

Project Pre-Check *FastPath* is a practice for stakeholder management. It is not a project management methodology, a change management methodology, a software development methodology or any of the other processes that are typically used to implement change. However, it does require that stakeholders assess the need for and determine which of these other processes will be required and used. It also relies on and interfaces easily with these other processes to implement stakeholder decisions.

Project Pre-Check views any change as a "business" initiative, demanding appropriate due diligence, prioritization mechanisms and appropriate justification and funding authorizations. That means that new technologies and enhancements to existing technology infrastructure should be managed by the same stakeholder practices that are used to deliver new products, manufacturing changes, mergers and acquisitions, new channels, new markets, business process enhancements, organizational restructurings, etc.

Why? Because, for a change to be successful, each of these ventures requires the active involvement and agreement of stakeholders throughout the enterprise. Project Pre-Check applies to all equally.

> *"The ability of the stakeholders to influence the final characteristics of the project's product and the final cost of the project is highest at the start and gets progressively lower as the project continues."*[13]
>
> *The Project Management Institute—A Guide to the Project Management Body of Knowledge*

Let's look at a real life project that turned out beautifully and see what lessons can be learned.

A Perfect Project

The Situation

This organization provided insurance and savings products to individual clients through over 100 Canadian offices coast to coast. There were 600 plus staff in these regional offices.

The office personnel responsible for administering clients' needs were faced with a motley array of mainframe, server and PC applications with widely different interfaces and formatting requirements, each application supported by its own 3-ring binder of procedures, entry protocols, codes and rules. The environment was a huge drain on productivity and quality and negatively impacted client service.

The Goal

The sponsor wanted to improve client service and the quality and productivity of the office staff by standardizing the look and feel of the interface and formatting requirements, providing online access to all support documentation, and ensuring that all new applications and future changes complied with the interface standards.

In addition, because of the number of staff affected, the new environment was to be sufficiently intuitive that it could be implemented effectively without formal training.

The Project

The organization reviewed available alternatives and launched the Interface Initiative, a tactical approach that would offer a standard PC based interface to the office staff (this was before the ubiquitous web browser). Behind the scenes this new interface would connect to and communicate with the existing application interfaces. Over time, those old interfaces would be eliminated and the new interface would communicate directly to the required function. This approach was seen as the best way to accelerate project delivery with manageable risks.

In addition, the project (known internally as "In Your Face") contracted with a recognized expert on usability and interface design and trained project staff to help address the no training target. All the content of the manuals would also be converted and accessible through context sensitive online help.

The project was sponsored by the VP Administration to whom the majority of the affected staff reported, through office managers and regional managers. The PM assigned had a proven track record as team builder and project manager. Project staff covered all the skills necessary to deal with the diverse set of technologies. An advisory council including ten office managers was formed to provide direction and feedback as design and development progressed and to liaise with their counterparts across the country.

The Results

The project was hugely successful. It won the Conference Board of Canada's ITX award for technology innovation and excellence. The ROI calculated by the (ITX Award) judging committee was about 800%.

The project delivered on plan and within budget and the high quality of the implementation was confirmed by the standing ovations the PM received as she travelled across the country introducing the new environment. When was the last time you received a standing ovation from your stakeholders??? Let me know!

How a Great PM Helped

The project's sponsor had a number of factors in his favour. The majority of the staff affected by the change was in his own organization so he didn't have to worry about the commitment of peers who might have had their own ideas about approach and priorities. There were no new technologies involved so the risks were largely known and manageable.

However, the great PM the sponsor and his IT counterpart selected turned what probably would have been a good result into a great result. His great PM:

- Helped mold the stakeholder group into a cohesive and productive force. The amount of collaboration was inspiring.
- Leveraged the stakeholders' authority to get the right resources at the right time.
- Clarified and quantified the project's goals and measured project performance against those goals on an ongoing basis.
- Communicated upwards, downwards and sideways continuously, in whatever form was necessary to gain understanding and agreement including one on one, group sessions, presentations, demos, email, written reports, coffee klatches . . .
- Used the project management, software development and management of change methodologies that existed within the organization and adapted them to suite the project's needs.
- Developed and managed a risk plan that resulted in some pragmatic changes in approach to reduce risk and improve quality and value
- Developed the application in phases and rolled out in stages to reduce risk and accelerate business value.
- Used industry best practices including usability, prototyping, partitioning and reuse to address the stakeholders' needs and ensure effective ongoing support of the delivered solution.

FastPath generally does not require stakeholders to spend more time, effort and energy on a change. However, it does require that the emphasis is placed at the front end, where involvement will have the greatest impact. It enables conscious, overt decisions about key factors that will influence how long the project will take, how much it will cost, what kind of quality will be delivered and what benefits will be delivered, when and to whom.

- It allows stakeholders to control the size and shape of a planned change based on identified risks and rewards, driven by anticipated business value.

Overview

o It integrates simply and effectively with other commercial and home grown project and change management and solution development methodologies.

o It can be applied effectively in the pre-launch stages, as part of or immediately after project initiation, up until solution delivery and on a release basis.

FastPath is also designed to change over time. The process allows each organization to adapt the best practices and Decision Areas to their unique circumstances, industry, culture and philosophies, up front and on an ongoing basis.

> *"It is our choices that show what we truly are,*
> *far more than our abilities."*
>
> *J.K. Rowling*

Project Pre-Check Building Blocks

Project Pre-Check is based on three fundamental building blocks: Stakeholders, Processes and the Decision Framework. Together, these building blocks provide the foundation for stakeholders to manage business and technology change successfully. *FastPath* shares these building blocks as well.

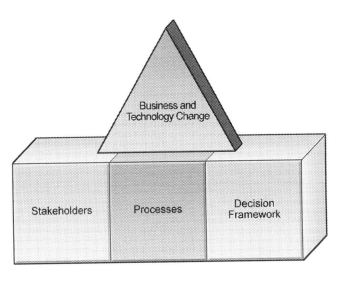

Figure 1—Project Pre-Check Building Blocks

Stakeholders are the decision makers: sponsors, change agents, targets and champions. Processes guide the decision makers to ensure that key steps are completed. The Decision Framework ensures stakeholders consider all aspects of the change, the environment within which the change is being implemented, the assets that can be leveraged or that will be affected and the way the project will be conducted. It focuses the stakeholders' decision making efforts on best practice areas that have helped other projects deliver successfully. Each building block is described in more detail below.

Stakeholders

Bralek's Rule for Success: Trust only those who stand to lose as much as you do when things go wrong

The Project Management Institute's Guide to the Project Management Body of Knowledge defines stakeholders as "individuals and organizations that are actively involved in the project, or whose interests may be affected as a result of project execution or project completion. They may also exert influence over the project's objectives and outcomes."[14]

Project Pre-Check takes a somewhat narrower view of the term "stakeholder". From the Project Pre-Check perspective, stakeholders are the influencers and decision makers. With the exception of the champion role, stakeholders are managers who have the organizational authority to allocate resources (people, money, services) and set priorities for their own organizations in support of a change. Champions, on the other hand, are selected for their ability to influence the behaviour of those who must change to support a successful implementation.

This narrower view is not meant to diminish or denigrate, in any way, the contributions of so many managers and staff whose efforts are essential to a successful change effort. It simply reflects Project Pre-Check's fundamental focus on helping the change decision makers do their jobs more effectively.

The rationale for this emphasis on decision maker is reinforced by the views of John Kotter, a professor at the Harvard Business School and the author of numerous books on corporate culture, change and leadership. Kotter's perspectives are summarized in an interview published in CIO Insight magazine under the heading "What are the conditions that mitigate against that (the change) succeeding?"

"I've seen too many technology projects get dumped on project teams and task forces that simply don't have enough clout, enough credibility, connections, you name it, to be able to do a difficult job, and so, surprise, surprise, they start getting frustrated and the powerful people in the company just ignore them or do what they want to do anyway.

Maybe it's an enterprise wide project, and the team that ought to be driving it is the CEO, his IT guy and a couple of others. And do they drive it? No. They give it to some group of suckers who kind of dribble into their meetings every once in a while to report. Well, it doesn't work.

Also, on a lot of the IT projects I've known, if you go up to the typical line manager and say to him, 'You've got this big thing going on here. What's the vision? Paint a picture for me. How's the company going to be different in 18 months when this is all done?' They can't even see it. So of course they haven't bought into it. And if they haven't bought into it, are they going to cooperate?"[15]

John Kotter—From CIO Insight Magazine

Project Pre-Check *FastPath* is targeted at project managers, to help them build the effective stakeholder group needed to manage and deliver successfully major business and technology change. It is their path to success. It demands active involvement very early in the change life cycle and requires ongoing agreement on vital decisions throughout the project.

As John Kotter and Dan Cohen suggest in their book, The Heart of Change, ".. . . the central issue is never strategy, structure, culture, or systems. All those elements, and others, are important. But the core of the matter is always about changing the behavior of people."[16] The stakeholders, in Project Pre-Check, are the agents for the behavioural change needed for project success.

The most critical factor in delivering change successfully is the active involvement and support of the right people. Kotter and Cohen claim "Large-scale change does not happen well without a powerful guiding force."[17] In Project Pre-Check, the stakeholders are that guiding force. Project Pre-Check mandates the commitment and agreement of stakeholders in a broad range of Decision Areas throughout the entire change life cycle.

In Project Pre-Check, there are four stakeholder roles as shown in Figure 2: sponsor, change agent, target and champion.

Sponsor

A Sponsor is a manager who:

o   Legitimizes the change
o   Has the economic, logistical or political power to make the change happen
o   Is responsible for decisions on the five w's—who, when, what, where, why
o   Is responsible for cost & benefit delivery

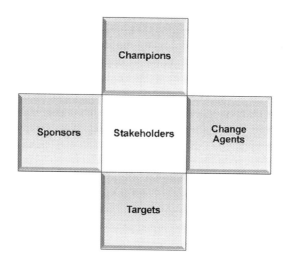

Figure 2—Stakeholder Roles

Change Agent

A Change Agent is a manager who is responsible to the sponsor for implementing the change. In many cases, the Change Agent role is filled by a project manager. However, that isn't always the case. Often, line managers wear the Change Agent hat in addition to their management duties. In either situation, authority is usually focused on determining "how" to deliver according to the sponsor's mandate. Change agents often do not have responsibilities beyond project completion.

The skills and motivation of change agents are key contributors to successful implementation. Managing the transition from the old state to the new state is a challenging task. It requires highly skilled people to guide the organization through the uncertainty of change in order to successfully achieve the organization's objectives. Organizations which select change agents based on who is available or expendable may be making a lethal error. Successful implementation requires change agents who have a formidable range of skills and motivation.

Target

A Target is a manager who directs individuals or groups who must actually change the way they operate and behave for a change to be successful. That change invariably leads to resistance, which is both natural and inevitable. The target stakeholders' role in developing strategies and tactics to anticipate and overcome the sources of resistance is critical to successful change implementation and effective ongoing operation.

Targets are usually responsible for the successful operation of their unit after implementation and need to steer the change throughout their organization to achieve a successful result. Targets include managers of departments within the initiating organization, such as marketing, sales, operations, manufacturing, information technology, finance, human resources and others. Targets can also include managers who are external to the organization initiating the change, such as customers, partners and distributors.

Champion

A Champion is a manager or senior staff member who enthusiastically supports the change and has the power, influence and respect necessary to help bring about the necessary behavioural changes in the managers and staff that are affected by the change.

For example, using a highly respected and highly successful sales person in the champion role to promote the introduction of a new product can help other sales staff overcome concerns they may have over the affect the new product will have on their clients, their performance and their income. The selected champion must be a firm believer in the value of the new product and must be able to converse effectively and convincingly with the other sales staff for the role to have value.

One or more champions can be selected to demonstrate and promote the new knowledge, skills and behaviours that will be required. That leadership and insight will help them explain and model the new skills and behaviours to target groups that will be faced with significant change, reducing resistance and accelerating the change process. Champions should be acknowledged innovators or early adopters who will readily rise to the change challenge, provide real and constructive feedback to the change's other stakeholders and be an inspiration to those that follow.

Changes in Perspective

Project Pre-Check may require you to alter your perspective on how changes are managed. It may also require other stakeholders to unlearn past beliefs and behaviours. For example:

o   Is the sponsor a manager who launches an initiative, passes it on to a project team and goes on to the next important challenge? Project Pre-Check requires that this individual stay focused on the change at hand, engage with the other stakeholders as a sponsor to wrestle with a myriad of decisions and ensure the change delivers the value the organization expects.

o   Are the targets managers who have sat back and watched as a project team struggled to understand and deliver a solution that met the needs of clients and staff in their organizations? Project Pre-Check needs them to collaborate with other stakeholders to ensure the project meets or exceeds your organization's requirements.

o   Are you a project manager who has tried to assume overall responsibility for a change or, conversely, looked to the sponsor for all decisions? Project Pre-Check expects you to join with the other stakeholders as a change agent to make the decisions required to maximize the value delivered by the project.

"Example is not the main thing in influencing others. It is the only thing."

Albert Schweitzer—Nobel Peace Prize winner

Project Pre-Check FastPath Process

Booker's Law: An ounce of application is worth a ton of abstraction.

A process is a collection of related activities that produce something of value to the organization, its stakeholders or its customers. In the case of Project Pre-Check, the thing of value is a successfully implemented change. The **FastPath** processes provide project managers with a road map to achieve that result.

Figure 3—Project Pre-Check FastPath Process

The Project Pre-Check **FastPath** process includes five stages to build and leverage the capability of the stakeholder group from project inception to completion: identify stakeholders, engage stakeholders, assess decision areas, monitor agreement and guide completion. These stages ensure the right players are involved, apply consistent rigor to the management of a change and make the decisions that need to be made to achieve success.

The stages incorporate the four key Project Pre-Check principles: accountability, relevancy, stakeholder agreement and integration.

- o Accountability places the mandate on one, and only one, stakeholder to provide guidance and direction to the stakeholder group on the relevancy and direction of a given Decision Area.
- o Relevance specifies whether a given Decision Area provides value to or is impacted by the planned change.
- o Stakeholder agreement tracks and measures the degree of stakeholder commitment to key project decisions. The success of a change initiative is predicated on stakeholders being accountable for the decisions that determine what benefits the change will deliver, when the benefits will be realized, how much the change will cost, what kind of functionality and quality will be delivered, who will need to change the way they perform for the change to be successful and what support they will be given to make those changes. Having all stakeholders on side with the decisions that shape a change offers the very best chance for success.
- o Integration leverages the in-place project planning and control mechanisms to make sure action items identified by the stakeholders are planned, resources are allocated, and progress is tracked to ensure completion. Integration relieves the stakeholders of the planning and control burden and allows them to focus on the critical decision making role.

Identify Stakeholders

Determining who the stakeholders are for a given project can be a most challenging task. The fundamental question that should be asked is who manages the organizations initiating or affected by the planned change and, within that group, who has the knowledge, ability and authority to make informed decisions regarding the impact of a planned change on the operations of those organizations.

Engage Stakeholders

This stage leverages stakeholder commitment at project start-up to agree on expected project outcomes, assess project impact on each Decision Area and agree on the contribution various Decision Areas can make to the success of the project.

The Engage Stakeholders stage is an ongoing exercise that establishes each stakeholder's need to be involved, their role in a project and their commitment to that role. It then facilitates and confirms their continuing involvement throughout the course of the project.

Assess Decision Areas

The Assess Decision Areas stage establishes stakeholder views on the relevancy of and accountability for each the 50 Decision Areas in the *FastPath* Decision Framework. It also identifies any additional Decision Areas (called Write-ins) that need to be considered over the course of the project. As well, it assesses the level of agreement on decisions taken and facilitates collaboration and consensus where gaps and variances exist

Monitor Agreement

Over the course of a project, things change. Required functionality evolves. Expectations regarding quality become clearer. Estimates and schedules change. People come and go, including stakeholders and project staff. Priorities change. The outside world imposes its will all too often in the form of new competitors, new products, legislation, political and environmental upheavals.

The Monitor Agreement stage is concerned with controlling scope, risk and organizational impact to deliver expected results. It provides the stakeholders with a framework for an ongoing review and management of decisions made and yet to be made in light of this changing milieu. It enables stakeholders to revisit decisions on Decision Area relevancy, accountability and levels of agreement, and to initiate corrective actions if needed, to balance costs, risks, time and value.

Guide Completion

Wrapping up a project is a journey over time. Certainly, at some point, one will be able to say "Ah ha! The project is finally done! Hooray!" But before that point is reached, a number of decisions need to be taken by the stakeholders: Have we delivered what we wanted to, for the price and in the timeframe required? What additional steps are needed to bank the benefits, to grow the benefits? Are there other features and functions that could be delivered that would provide incremental value? Are there other competing priorities that require the resources? What resources should be released and when? What worked, what didn't and what lessons learned could we leave for others? Etc.

Answers to these and other questions can extend the project, cause the project to be terminated prematurely, spawn additional initiatives, cause organization priorities to shift and any number of other outcomes.

The Guide Completion stage offers stakeholders a mechanism for rational deliberations and decision making and effective project closure. In addition, the Guide Completion stage, like the other stages, should iterate until the conditions are deemed appropriate to wrap things up.

The *FastPath* stages are primarily for project managers, to facilitate the decision makers on the change. They govern the scope and deliberations of the stakeholder group, they foster active involvement in and agreement with the decisions required and made and they help stakeholders apply the insight and value from proven industry and organizational best practices to ensure project success.

The *FastPath* process interacts effectively with other in place project management, system development and change management methodologies and includes the capability to shape and customize *FastPath* to the needs of an organization, enterprise or industry.

"Knowing is not enough; we must apply. Willing is not enough; we must do."

Goethe, German writer and scientist.

Decision Framework

How many decisions do stakeholders make guiding a change through to completion? What kind of decisions do they make? When do they need to make them?

Project Pre-Check provides structure to that decision making process through the Decision Framework. It includes four Domains; Change, Environment, Assets and Project as shown in Figure 4. Those four Domains contain eighteen Factors and 125 Decision Areas that need to be considered for each and every change. The *FastPath* Decision Framework focuses on the 50 Decision Areas most frequently relevant to project success.

The Decision Framework helps stakeholders maintain a broad perspective of the planned change, including business and technology, people and processes, internal and external organizations, short term and long term views.

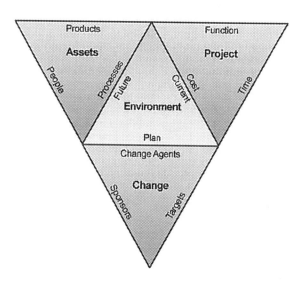

Figure 4—Project Pre-Check Domains

The Decision Framework also provides the means to rationalize thousands of best practices from a multitude of diverse sources into the Decision Areas that form the basis for Project Pre-Check decision making.

It is also the mechanism that is used to support the concept of "Write-ins"—Decision Areas that project stakeholders can add to their decision making deliberations to cover the unique needs of the project, the organization, industry or culture. The "Write-ins" can be applicable to the scope of one project or can be added to the Decision Area catalogue for use by other projects and stakeholder groups.

The Decision Framework Domains and Factors are reviewed in the following pages. The Decision Areas are addressed in detail in Part V. The Factors included in the *FastPath* framework are in bold in the diagram below. The excluded Factors are shaded out.

Change Domain and Factors

The Change Domain focuses on stakeholder decisions regarding the specifics of the change itself. It includes the dimensions of the change initiative, the quality needs and expectations that must be satisfied, the investment decisions behind the change and a review of stakeholder capabilities. It includes the following Factors as shown in Figure 5:

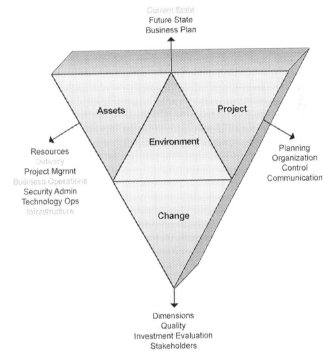

Figure 5—Decision Framework Domains & Factors

o Dimensions define the breadth and depth of the opportunity or problem being addressed.

o Quality defines the expected levels of performance for the planned change.

o Investment evaluation deals with the rationale for the change, risk levels and relative priority.

o The Stakeholders Factor considers roles and responsibilities, commitment levels and capabilities.

These Factors were derived from an analysis of practices that have been proven to contribute much to project success, including:

o Executive sponsorship.
o Well defined objectives.
o Business cases.

    o   Addressing quality expectations up front.

    o   Rapid, incremental, delivery.

    o   Collaboration, communication and agreement among stakeholders.

    o   Understanding stakeholder capabilities.

Environment Domain and Factors

The Environment Domain embodies the operational context of the enterprise: what exists today, what is the future vision and how does the organization plan to get there. *FastPath* includes the following Factors as shown in Figure 5:

    o   Future state addresses what is planned for the future that may change as a result of the planned change and the impact on inter-relationships.

    o   Business plan addresses the impact of the planned change on the overall business direction and priorities.

Proven best practices that contributed to the development of the Environment Domain Factors and Decision Areas include:

    o   Well-articulated mission, vision, goals and strategies.

    o   Integration of business and technology plans to support the organization's strategies and achieve stated goals.

    o   Architecture—mapping out a desired future state and how to get there.

    o   Understanding the impact of a change from the outside in—from clients, providers, partners, through service organizations to the core business operations.

Assets Domain and Factors

The Assets Domain addresses stakeholder decisions regarding people, processes and products. It is like the file cabinet to the project. It identifies the assets that can be leveraged, or that need to be changed, added or deleted to support a planned project. *FastPath* includes the following Factors as shown in Figure 5:

    o   The Resources Factor includes internal and external human resources and physical plant.

o   The Project Management Factor addresses the project management processes that are required or available and can be leveraged to manage the planned change.

o   Security Administration focuses on the value from and impact on the key aspects of the overall security program.

o   Technology Operations encompasses the technology management processes that will be required to support the planned change during its development and on and after implementation.

The following best practices contributed to the identification and development of the Assets Domain Factors:

o   The right people, with the right skills, in the right place, at the right time.

o   Team formation and development.

o   Robust business and technology processes.

o   Proven project and change management practices.

o   Building quality in from the start.

o   Tackling risks according to potential impact.

o   Active risk, issue and change management.

o   Leveraging and reusing assets, including best practices.

Project Domain and Factors

The Project Domain addresses stakeholder decisions on how the project or projects that deliver the change will be conducted. It specifies key project management parameters that need to be considered by the stakeholders to ensure they are able to make effective decisions on time, cost and function delivery. It includes the following Factors as shown in Figure 5:

o   The Planning Factor addresses the application of relevant best practices identified in the Assets Domain to deal with the unique needs of the planned change.

o   Organization includes the project organization structures, roles and associated resources.

o   Control addresses the mechanisms that help stakeholders shape and guide the project to achieve the desired goals.

o Communication deals with the ongoing practices and decisions that will ensure all participants in the change are fully involved and informed in a timely manner concerning the progress of the planned change.

The Project Domain Factors outlined above were founded on a number of proven best practices, including the following:

o Actively managing the change, not just the project.
o Well-articulated goals and objectives.
o Always considering alternatives.
o Keeping implementations small.
o Pragmatic use of metrics and measurement.
o Rigorous, defensible estimates.
o Incremental delivery
o Delivering solutions that are "good enough".
o Stakeholder commitment to the change and active involvement throughout.

Managing Decision Areas

There is an incredible variety of expertise and intelligence on how to ensure successful project delivery. The challenge, of course, is how to tap that wealth of knowledge to deliver immediate, cost-effective value to a project or an organization.

The best practices and derived Decision Areas included in this book were selected based on their ability to address the issues that frequently plague major change initiatives, including cost overruns, schedule slippage, quality problems, missed benefit targets, functionality and usability gaps, security gaps, lack of responsiveness (takes too long to deliver), performance issues and user acceptance issues.

Nicholas Carr, author of Does IT Matter? and former executive editor of the Harvard Business Review, presented the following lessons in his Rough Type blog. While the article deals with software projects, the lessons apply equally to other types of major business and technology change ventures.

"First is that the bigger the software project, the more likely it is to collapse under its own weight. $100 million seems like the line beyond which failure is almost assured.

Second is that you should always create software to solve the day-to-day problems faced by the actual users, not to meet big, abstract organizational challenges. Solve enough little problems and the big ones take care of themselves. Fail to make users lives easier, and they'll simply bypass the system (and never trust anything you do ever again).

Third and finally, you should never give a software project a catchy codename. For Ford, it was Project Everest; for McDonald's, it was Project Innovate; for Lloyd's, it was Project Kinnect. If you're about to launch an IT initiative big enough to warrant its own name, you should probably make sure you have a really good golden parachute."[18]

So, what is a Decision Area and how does it add value?

Definitions: Best Practice and Decision Area

In Project Pre-Check parlance, a best practice is a technique or approach to solving a problem or delivering a solution that has the following attributes:

o   It has been tried and proven, internally or externally
o   It contributes to productivity, quality, cost, risk or responsiveness in a positive manner.
o   It is acknowledged as a "current best practice" by a number of peers
o   It is documented, communicated and has been applied successfully
o   It is a special insight on a topic, but not necessarily the last word
o   It focuses on how, on real experience
o   It is relevant and adds value to the organization

Best practices can address a wide variety of disciplines including business, technology, human resources, organization structure and operation, contractual matters, legislative and economic concerns and more. They are concerned with specifying how a specific end is to be accomplished.

Overview

By their very nature, best practices are transitory. They may or may not be related to specific technology but they are capable of being superseded by new "current best practices" as business insights, technology changes and project needs dictate.

Where a best practice is procedural in nature (it usually includes "how to" instructions), a Decision Area identifies "what" and provides the label to the procedure. A Decision Area is simply a best practice raised to a higher level. For example, while there are numerous risk management practices available, effective risk management of some sort is generally essential for project success. The Decision Area in this situation is risk management. The best practice is the specific risk management practice selected. Decision Areas represent any of the following:

o Things that need to be defined (e.g. Goals, success criteria)
o Impacts a project will have that need to be assessed (e.g. Organization mission, vision, strategies)
o Practices that need to be considered (e.g. Prototyping, estimating)
o Processes that are in place that may be affected by a change (e.g. Service desk, security administration)
o Solution attributes that may need to be established for a change (e.g. Business transaction volumes, performance expectations)
o Factors that could have a significant influence on the kinds of alternatives considered to solve the problem or address the need (e.g. usability, flexibility).

Mapping Best Practices to Decision Areas

The process for selecting best practices and translating them into Decision Areas takes the following path, as shown in Figure 6:

o Best practices that appear to offer significant potential for avoiding project pitfalls are analyzed to identify the relevant Domain and Factor to rationalize multiple, overlapping best practices to a common framework.
o If the best practice is already addressed by a Decision Area, the Decision Area definition is checked to ensure it encompasses the best practice fully and is revised if necessary.
o If the best practice is not covered by an existing Decision Area, a new one is created and added to the appropriate Factor and Domain. This is an ongoing process to ensure

Project Pre-Check reflects the best intelligence and latest insights into the practices that contribute most to successful change.

o   Organizations can tailor the Decision Areas to their ongoing needs to cover project, company, industry or country specific concerns and incorporate new learnings from internal and external sources. In fact, the Project Pre-Check processes support the concept of Write-ins, which allows stakeholders to add new Decision Areas dynamically, as the need arises. Existing Decision Areas that are not relevant to a particular organization can be deleted or ignored as appropriate.

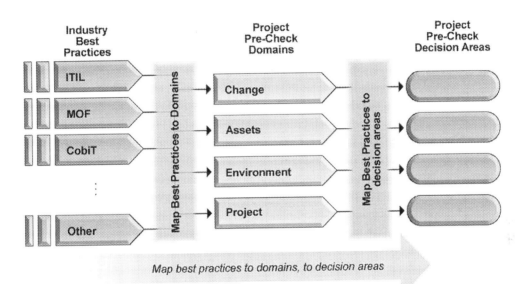

Figure 6—Mapping Best Practices to Decision Areas

Which Practice to Use

Project Pre-Check is built on years of review and experience with a variety of best practices. It does not dictate which specific best practice implementation you should apply. Instead, it identifies areas of emphasis that industry experts and experience suggest can help organizations deliver responsive, cost-effective, quality solutions consistently over time. The selection of which Decision Area to focus on and which specific best practices to use will depend on the project circumstances, the capabilities of the organization and the priorities of the risks and issues that need to be addressed.

Using Decision Areas

Project Pre-Check uses the Decision Framework to screen a multitude of best practices, filter out the duplicate or conflicting ones and derive or amend appropriate Decision Areas. This ensures a manageable number of Decision Areas that stakeholders can consider to determine areas that may be impacted by a change or that may contribute to delivering a change successfully. Decision Areas enable stakeholders to quickly shape the planned change, identify areas of impact and take advantage of opportunities that can quickly and effectively add value to a given initiative.

The chosen Decision Areas are intended to be sufficiently general that they can be understood by the vast majority of stakeholders regardless of their backgrounds and expertise and can apply to the majority of major changes. Each Decision Area is covered in more detail in Part V.

"The important thing is not to stop questioning."

Albert Einstein

One of the following decisions should be made by the stakeholders for each of the 125 plus Decision Areas. The items in brackets are examples of specific Decision Areas covered later in the book.

o   The Decision Area has no relevance to this project (e.g. Using a prescribed methodology, succession planning)

o   The Decision Area is relevant to the project (e.g. Team formation, regulatory compliance)

o   We don't yet know whether the Decision Area is involved with or affected by the planned change (e.g. System management capacity, phasing options)

"It is better to know some of the questions than all of the answers."

James Thurber
American author and cartoonist

The Decision Areas included in this book provide a comprehensive starting point for any planned change and provide beneficial insight for the stakeholders. The "Don't Yet Know" response is a very valid initial stance that is used to drive follow-on activity. However, the earlier the "don't know" responses disappear, the more comfortable stakeholders will be and the less chance surprises will occur.

The Decision Framework Value Added

The Decision Framework, with its four Domains, 18 Factors, 125 Decision Areas and Write-in capability is designed to encompass every element that any project, anywhere, will need to consider.

How, you might ask, does this one model cover every concern, in every project, in every organization, industry and culture? Take a look at the Project Pre-Check Domain structure:

o   The Change Domain includes Decision Areas that describe the planned change in all its breadth and depth.
o   The Environment Domain includes the Decision Areas that relate to the external environment and the context of the change.
o   The Assets Domain includes the Decision Areas that can provide value to the change or that are impacted by the change.
o   The Project Domain includes the Decision Areas that deal with delivery of the planned change.

If these Domains and Decision Areas don't cover a Factor, then the Write-in facility can be used to ensure it's captured and addressed.

The following diagram (Figure 7) illustrates the breadth of the Decision Framework. The highlighted factors and Decision Areas represent the 50 Decision Areas in *FastPath*.

At recent seminars on project auditing, the audiences were asked to examine the diagram, identify Decision Areas that should be included and to identify Decision Areas that would have no relevance to an organization in the foreseeable future. The question was met with a collective shrug in each session.

Certainly, there are Decision Areas relevant to specific companies, industries or cultures that can be added to provide incremental value. As certain, some of the existing Decision Areas will provide limited value depending on the circumstances. Project Pre-Check is designed to adapt as needed to help stakeholders guide each change to a successful conclusion.

Overview

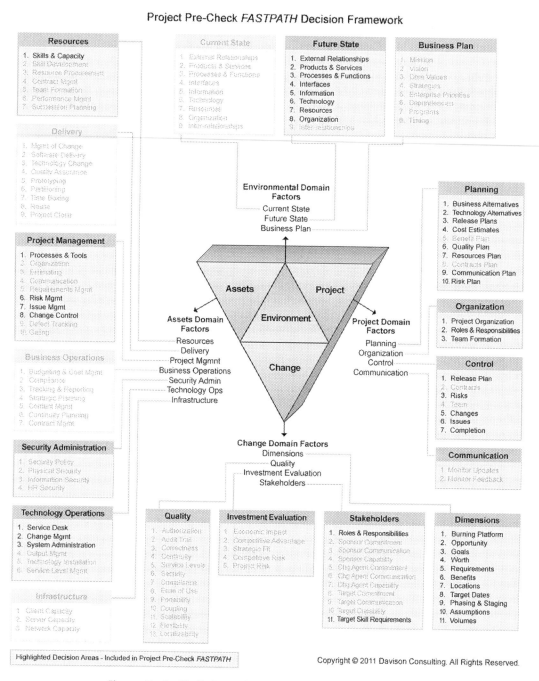

Project Pre-Check *FASTPATH* Decision Framework

Resources
1. Skills & Capacity
2. Skill Development
3. Resource Procurement
4. Contract Mgmt
5. Team Formation
6. Performance Mgmt
7. Succession Planning

Current State
1. External Relationships
2. Products & Services
3. Processes & Functions
4. Interfaces
5. Information
6. Technology
7. Resources
8. Organization
9. Inter-relationships

Future State
1. External Relationships
2. Products & Services
3. Processes & Functions
4. Interfaces
5. Information
6. Technology
7. Resources
8. Organization
9. Inter-relationships

Business Plan
1. Mission
2. Vision
3. Core Values
4. Strategies
5. Enterprise Priorities
6. Dependencies
7. Programs
8. Timing

Delivery
1. Mgmt of Change
2. Software Delivery
3. Technology Change
4. Quality Assurance
5. Prototyping
6. Partitioning
7. Time Boxing
8. Reuse
9. Project Close

Project Management
1. Processes & Tools
2. Organization
3. Estimating
4. Communication
5. Requirements Mgmt
6. Risk Mgmt
7. Issue Mgmt
8. Change Control
9. Defect Tracking
10. Gating

Business Operations
1. Budgeting & Cost Mgmt
2. Compliance
3. Tracking & Reporting
4. Strategic Planning
5. Content Mgmt
6. Continuity Planning
7. Contract Mgmt

Security Administration
1. Security Policy
2. Physical Security
3. Information Security
4. HR Security

Technology Operations
1. Service Desk
2. Change Mgmt
3. System Administration
4. Output Mgmt
5. Technology Installation
6. Service Level Mgmt

Infrastructure
1. Client Capacity
2. Server Capacity
3. Network Capacity

Environmental Domain Factors
- Current State
- Future State
- Business Plan

Assets

Project

Environment

Change

Assets Domain Factors
- Resources
- Delivery
- Project Mgmnt
- Business Operations
- Security Admin
- Technology Ops
- Infrastructure

Project Domain Factors
- Planning
- Organization
- Control
- Communication

Change Domain Factors
- Dimensions
- Quality
- Investment Evaluation
- Stakeholders

Planning
1. Business Alternatives
2. Technology Alternatives
3. Release Plans
4. Cost Estimates
5. Benefit Plan
6. Quality Plan
7. Resources Plan
8. Contracts Plan
9. Communication Plan
10. Risk Plan

Organization
1. Project Organization
2. Roles & Responsibilities
3. Team Formation

Control
1. Release Plan
2. Contracts
3. Risks
4. Team
5. Changes
6. Issues
7. Completion

Communication
1. Monitor Updates
2. Monitor Feedback

Quality
1. Authorization
2. Audit Trail
3. Correctness
4. Continuity
5. Service Levels
6. Security
7. Compliance
8. Ease of Use
9. Portability
10. Coupling
11. Scalability
12. Flexibility
13. Localizability

Investment Evaluation
1. Economic Impact
2. Competitive Advantage
3. Strategic Fit
4. Competitive Risk
5. Project Risk

Stakeholders
1. Roles & Responsibilities
2. Sponsor Commitment
3. Sponsor Communication
4. Sponsor Capability
5. Chg Agent Commitment
6. Chg Agent Communication
7. Chg Agent Capability
8. Target Commitment
9. Target Communication
10. Target Capability
11. Target Skill Requirements

Dimensions
1. Burning Platform
2. Opportunity
3. Goals
4. Worth
5. Requirements
6. Benefits
7. Locations
8. Target Dates
9. Phasing & Staging
10. Assumptions
11. Volumes

Highlighted Decision Areas - Included in Project Pre-Check *FASTPATH*

Figure 7—FastPath Domains, Factors and Decision Areas

Using Project Pre-Check FastPath

Brien's First Law: At some time in the life cycle of virtually every organization, its ability to succeed in spite of itself runs out.

The Project Pre-Check building blocks described above provide the foundation for stakeholders to manage change successfully. The diagram below summarizes how the building blocks work together to achieve that goal.

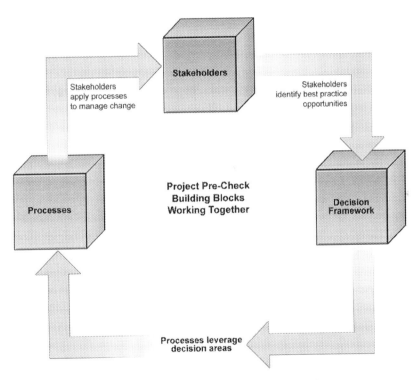

Figure 8—Building Blocks Working Together

There are a couple of options for leveraging *FastPath* to get up and running quickly and effectively. The easiest, and the recommended approach, is the Fast Start approach using the *FastPath* process on an existing, in-progress project. The options for delivering Project Pre-Check value are described below.

"What you do speaks so loudly that I cannot hear what you say."

Ralph Waldo Emerson

Fast Start—For Active Projects

If the target project is already underway—defined, estimated, funded, resourced and in progress—the *FastPath* process can be used as a starting point to identify and engage the stakeholders and establish the degree of stakeholder agreement on the key Decision Areas. The first three stages will identify gaps and areas of stakeholder concern that need to be rectified quickly to minimize project risk.

o The Identify Stakeholders stage confirms the key players who should be involved in the *FastPath* process.

o The Engage Stakeholders stage brings all stakeholders up to speed on their role in the project and the *FastPath* process.

o The Assess Decision Areas stage involves sessions with each of the stakeholders, by interview or questionnaire or by group collaboration. Each stakeholder reviews the *FastPath* Decision Areas, records their own views on accountability, relevancy and their level of agreement an decisions already made. The results are consolidated for review and identification of action items.

o The Monitor Agreement stage leverages the insights gained to reach agreement and resolve gaps and areas of divergence on Decision Area accountability, relevancy and levels of agreement and monitors Decision Area issues that need further consideration as the project progresses.

o Finally, the Guide Completion stage can help stakeholders steer the project to a successful final resting place.

The action items should be integrated into the existing project plan for resolution and can feed into the *FastPath* Monitor Agreement stage for the duration of the project. The *FastPath* process can also be used on a phase by phase and release by release basis.

Remember, the whole *FastPath* process, and each stage within the process, should be iterated as needed. If a new decision area is revealed or a change to the project has caused some stakeholders to have second thoughts about a previous decision, redo the Assess Decision Areas stage for those issues. If a stakeholder leaves, changes jobs or becomes incapacitated, revisit the Identify Stakeholders and Engage Stakeholders stages again.

Optimum Value—For New Projects

Putting Project Pre-Check *FastPath* to work at the start of a new project will deliver maximum value to that project. It's as simple as applying the *FastPath* stages as described above. However, because the project is new, very few decisions will have been made. Consequently the Assess Decision Areas and Monitor Agreement stages should be iterated as needed until there is comprehensive agreement by all stakeholders on all relevant Decision Areas.

Maximum Impact—For All Projects

Project Pre-Check *FastPath* can be applied quickly and effectively one project at a time. However, there are significant organizational benefits if the insights and expertise from numerous projects and talented people can be aggregated and made available to others to improve project performance across the enterprise.

FastPath can be the cohesive force that brings the best practices together, rationalizes them using the Decision Framework and offers them to all enterprise changes through the process stages and Decision Areas.

To implement Project Pre-Check *FastPath* as a standard practice across a department, organization or enterprise, the steps are the same—apply *FastPath* to implement *FastPath*, as outlined above and detailed in the process chapters that follow.

Lead by Example—For All Projects

A good friend of mine, whose recent assignments have focused on rescuing troubled projects, was enthusiastic about the original Project Pre-Check and the proven practices it helps stakeholders access. However, he expressed reservations about his own ability to deliver value from Project Pre-Check where he's functioning as a newly appointed Project Director and the other stakeholders are screaming for results.

If you can't impose the use of Project Pre-Check on a project, be surreptitious! As a project manager, if you lack the clout to dictate practices, lead by example! Don't bother telling your stakeholder partners you're using something called *FastPath*. Simply use the building blocks to ask questions, solicit feedback and facilitate agreement.

Overview

The beauty of the Lead By Example approach is that you'll be doing what a capable, actively involved stakeholder should be doing anyway. Some of your peers will be impressed with your knowledge, insight and leadership and will want to participate. Others will be oblivious. No matter! Your actions will still make a significant contribution.

"There's only one corner of the universe you can be certain of improving, and that's your own self."

Aldous Huxley (1894-1963)
British writer, author of "Brave New World" Part II

PART III
Stakeholders

- o **Managing the Stakeholder Group**
- o **Coping with Stakeholder Change**
- o **Joining an In Flight Project**

While leadership is generally expected from the sponsorship role, in Project Pre-Check, all stakeholders, in all roles, must be leaders to ensure project success. According to the findings of James Kouzes and Barry Posner in their book, The Leadership Challenge, there are "five fundamental practices in exemplary leadership:

- o   Leaders inspire a shared vision. They envision the future.
- o   Leaders challenge the process. They search for opportunities to change the status quo. They experiment and take risks
- o   Leaders enable others to act. They foster collaboration and build spirited teams. They strengthen others.
- o   Leaders model the way. They set an example. They plan small wins.
- o   Leaders encourage the heart. They recognize contributions and celebrate accomplishments."[19]

These are the behaviours that all stakeholders need to model for a change to be successful. These are the behaviours that Project Pre-Check depends on to ensure the success of the change process.

Stephen Covey, author of The 7 Habits of Highly Effective People, sums up this leadership need in his new book, The 8th Habit: From Effectiveness to Greatness: "Find Your Voice and Inspire Others to Find Theirs"[20].

Managing the Stakeholder Group

Obviously, one person can fill one or more roles on a project. A sponsor must also "champion" the change and may be a target of the change (i.e. skill and behavioural changes are required personally or by staff in his or her organization). A sponsor may also be responsible for directly managing a part of the change as a change agent. Similarly, a manager who is a change target must also act as a sponsor and champion to his or her reports.

What's important is that the roles are understood and filled so there is at least one person whose primary role is sponsor, change agent, target and champion from the inception of the change through completion. A major change will typically have from five to ten stakeholders: preferably just one sponsor, one or two in the change agent role, three or more in the target role and one or more in the champion role. The actual number depends on the decision makers that need to be involved for the project to be a success. The diagram below (Figure 9) shows how the stakeholder group for a planned change can be formed to include the people in the various organizations involved with and affected by the change.

This is the group that will be charged with making the decisions needed to achieve the expected results.

The easiest way to identify the necessary participants in the stakeholder group is to use an organization chart to identify accountability and impact. It is preferable to have participants from similar management levels to cultivate collaboration. Also, with the exception of the champion, only one person per organization should be included.

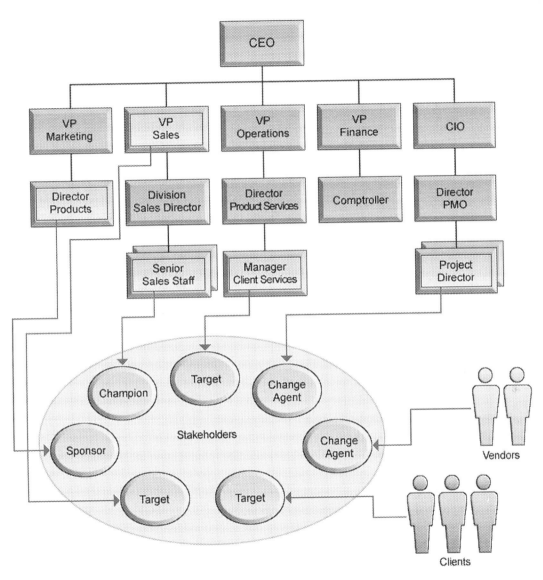

Figure 9—Creating the Stakeholder Group

For example, in Figure 9, the VP Sales is included instead of all the Division Sales Directors. Having both the VP Sales and one or more Sales Directors in the stakeholder group would be counter-productive. Notice too that the stakeholder group is represented as an oval, not a hierarchy—a collaborative organization rather than command and control. The rationale is this: in many projects the stakeholder group is made up of individuals from a number of different organizations that have competing and conflicting needs and priorities. Those conflicts can never be effectively resolved by edict. The solutions need to be considered, vetted, negotiated and embraced collaboratively.

The stakeholders will be self-sufficient as long as all players have the authority and capability to make decisions on behalf of the organizations they represent, all the necessary organizations are represented and the consensus process works. Where conflicts arise that cannot be resolved within the group, the issues should be quickly escalated up the organization until a decision can be reached. Escalation should never be construed as a sign of failure of the stakeholder group. Rather, it should always be positioned as an available and appropriate measure for resolving a deadlock.

How does the Project Pre-Check stakeholder group relate to traditional project steering committees? If the rationale for and participation in the two groups is similar, perhaps only one group is required. However, in many cases, a steering committee is formed to provide an oversight function rather than a decision-making role. In those cases, a steering committee may still be appropriate. Each organization will need to establish its own guidance within the corporate governance structure.

The above diagram also includes a client functioning as a target stakeholder in the stakeholder group. Often this arrangement is impractical but it can yield terrific results if appropriate for the change and the right client consents. Where direct client involvement is not possible, the manager responsible for the relationship with the client groups affected by the change should be a stakeholder operating as a target.

"You have brains in your head. You have feet in your shoes. You can steer yourself any direction you choose."

Theodor Seuss Geisel

Coping with Stakeholder Change

Sometimes you're lucky enough to be involved in a terrific project that's progressing well and being guided by an engaged, talented, committed stakeholder group. It's a great experience!

And then you're reminded of that old adage—the only constant is change. One of the stakeholders departs. It could be a move to another organization, or an internal reassignment, but suddenly there's a big hole to be filled. And it does have to be filled! Whether the departed stakeholder was a sponsor, change agent, target or champion, someone new will have to be brought in to fill the role.

I have seen projects try to get along without replacing a departed stakeholder. Big mistake! If there was a valid reason for that individual to be involved in the first place, then there's an equally valid reason to ensure a replacement is found and engaged.

When I run Project Pre-Check courses, one of the first things I do is ask the participants to identify one or two bad projects and one or two good projects they've had experience with and the factors that made them failures or successes. In one of those courses, a participant talked about a project he was on that had three different sponsors over a ten year period. The first sponsor initiated the project, a multi-million dollar affair, and provided reasonably good direction and guidance until her departure thirty months in. Her replacement, who should have filled the sponsor role, didn't care about the project and wasn't interested or involved. The other stakeholders soon recognized that the project wasn't going anywhere and lost interest as well. But for some reason the funding was still in place so the project director kept the project going. And going! And going!

After ten years, YES, TEN YEARS!!!, a new executive arrived on the scene, discovered the wayward project, still plugging along with the same project director and a team of fifteen, and pulled the plug. What a waste! What should have happened?

Stakeholders

An effective stakeholder group is more than the sum of its parts. Members share not only the specific knowledge they've acquired over the course of the project, they also share relationships they've built over time and the insights they've acquired from the project decisions that have been made. The challenge is to bring a replacement quickly up to speed as a fully functioning member of the stakeholder group. The following four steps will help realize that goal:

- o   Identify the New Stakeholder

    Identifying the new stakeholder may be easier said than done. If the departed stakeholder was a sponsor or target and the vacated position has been filled, the new manager is the obvious candidate. However, if the position has not been filled, finding someone to cover the sponsor or target role is critical. The obvious candidates can be found within the organization of the departed stakeholder, preferably the departed's immediate superior.

    If the departed stakeholder was a change agent or champion, then recruiting activity will be needed to find a permanent replacement. This is where things can get dicey, especially with regard to the change agent role! Is there someone who can fill in effectively on a temporary basis until a replacement is found? If not, there are two options; stumble along until the replacement is on board or, suspend the project.

- o   Engage the Stakeholder

    Having identified the replacement, it is crucial to the success of the project that the individual agrees to fulfill the required role (sponsor, target, change agent or champion) and commits to the ongoing oversight of the initiative as an active member of the stakeholder group.

    Recognize that the new stakeholder will have lots on their plate, especially a new sponsor or target. Not only do they have to assume the responsibilities involved in guiding the project, they may have a large organization to manage and untold other duties to understand and carry out. Getting the new member to commit formally to the new role is essential.

o   Integrate the Stakeholder

There is nothing more frustrating for a high performance team than to be sidetracked debating things already resolved or to drag along a new member who doesn't have the depth or breadth of understanding to participate effectively.

So, once the new stakeholder has been identified and engaged, it's time to bring them up to speed on all aspects of the project so that their deliberations and contributions going forward are in synch with the other members of the stakeholder group.

This is where something like Project Pre-Check's Decision Framework can provide huge value when it comes to accelerating and enhancing a new stakeholder's contribution to the project. It provides a vehicle for recording stakeholder decisions on what the project is trying to achieve and why, what's in and what's out in terms of scope, where everyone is in agreement with a decision and where there are still differences of opinion. The decision record will also point to the documentation and other sources that reflect the results of the stakeholder deliberations. That's a rich project history and a vital source of information for the new stakeholder.

Equally important, however, is the involvement of the other stakeholders in the new member's enlightenment. Don't just dump a load of documentation in the new stakeholder's lap. Get the other stakeholders involved, in one on one or small group sessions, to contribute their personal views on the project and the decision records. That will help build the relationships and reinforce the culture of the stakeholder group to enhance future group performance.

o   Operate As a Full Member of the Stakeholder Group

Once the new stakeholder has been identified, engaged and integrated into the operations of the stakeholder group, all that's left is to make the group work to guide the project to a successful conclusion. However, the existing stakeholders still need to keep an eye on how the new member is performing and take corrective action if issues arise. Remember, the new stakeholder is just getting up to speed and will be subject to a variety of distractions, especially if they're new to the organization. Continued vigilance is essential!

Joining an In Flight Project

There are lots of challenges taking on a new job, getting up to speed quickly, putting your stamp on your group and contributing effectively to the larger organization. Often, with that new job comes essential involvement in an in progress change initiative. Your new role could be as sponsor, target, change agent or champion. Regardless, it's an added responsibility that needs to be accommodated in your already busy calendar.

The three steps below are somewhat of a mirror image to the previous post but in this case, you're managing the process. If the other stakeholders involved have read my previous post and are carrying out those four steps, so much the better.

Here's your road map for getting up to speed quickly:

o   Identify the Stakeholders and Their Roles

> You need to know who the other project decision makers are and what roles they're playing—sponsor, target, change agent or champion. That includes you! The easiest way to find out is to ask the other stakeholders. You want to see consistency in the responses. If not, make a note to follow-up. Also, do some digging on your own: Who was responsible for launching the project? Who does the organization look to to deliver the change successfully? Which departments and which staff will be affected by the change? Are those groups represented in the stakeholder group? If you find issues or gaps in your digging, make a note to follow-up.

> This step is absolutely critical to get you up to speed quickly and contributing effectively but it's often overlooked. An example: a colleague of mine recently joined a project as a contract project manager. It involved the launch of a new financial product with a boat load of web development. She jumped into the project with both feet in response to some critical target dates. And then she ran into a number of conflicts. The VP of System Development, her boss, explained that he was the project sponsor and proceeded to dictate direction,

which was at odds with business expectations. The resource manager with whom she dealt to obtain project staff, issued orders regarding the functionality, look and feel of the application, again at odds with the business direction.

The project manager took a deep breath and stepped back. She asked the individuals involved who they thought the stakeholders were and what roles they were playing. She found considerable disagreement and pursued the issues until there was unanimity. As a result of her questioning, a number of changes were made to the makeup of the stakeholder group. The business VP responsible for the product launch became the sponsor. The System Development VP had no formal stakeholder role, nor did the resource manager. Two organizations affected by the project were not initially represented in the stakeholder group. Senior managers were appointed to fulfill their target roles. According to the project manager, once the stakeholder group was reformed, the difference in clarity on goals and directions, in the speed of decision making and the level of stakeholder commitment was palpable.

o   Review Decision Area Status

Now that you have a stakeholder group that covers all the bases, the next step is to discover what decisions have been made and agreed to and what decisions have yet to be addressed. The easiest way to cover this is through the use of a tool like the Project Pre-Check Diagnostic process. It uses as standard questionnaire covering the Decision Framework and ascertains the level of understanding and agreement for each stakeholder on each decision area. It can be completed in one on one or group discussions and cover all or a subset of the 125 Project Pre-Check or the 50 *FastPath* decision areas.

It doesn't matter what role you, the newcomer, are playing. It's as appropriate for a sponsor as it is for a change agent, target or champion. It's an incredibly powerful tool that will give you, as well as the other stakeholders, insights into gaps that need to be addressed and differences of opinion that need to be resolved. Also, it will give you access to the information sources that support the decisions made to allow you to bring your level of knowledge and understanding up to the level of the other stakeholders. No doubt, the other stakeholders will be impressed!

o   Operate As a Full Member of the Stakeholder Group

> Once the stakeholder group has been formed and the status and level of agreement on the required decisions have been established, all that's left is to join with the other stakeholders to deliver the project successfully. That includes resolving areas of stakeholder disagreement, tackling gaps that need attention and monitoring the impact on decisions already made as changes in scope and plan are approved over time. Here the results from the decision area assessment can serve as an ongoing scorecard to track decisions made and levels of stakeholder agreement.
>
> As the new player in the stakeholder group, you can take pride in your contributions, confirming the stakeholder group makeup or reshaping as needed, identifying the level of agreement with decisions made and determining areas needing attention. Not bad for a newcomer!

PART IV
Fastpath Process

o **Identify Stakeholders**
o **Engage Stakeholders**
o **Assess Decision Areas**
o **Monitor Agreement**
o **Guide Completion**
o **Managing the Process**

The Project Pre-Check *FastPath* process is designed primarily for those filling the change agent role, usually project managers. The process is easy to learn, simple to apply and quick to generate value. However, it may require a shift in your approach to project and change management. *FastPath* requires you, as a change agent, to put considerable effort into facilitating stakeholder decision making, recognizing that the investment in time and energy will pay off handsomely down the line.

"Why not go out on a limb? Isn't that where the fruit is?"

Frank Scully

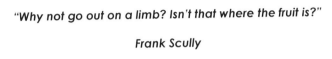

Figure 10—FastPath Process

Also, *FastPath* will almost certainly involve a change in the other stakeholders' views of what their responsibilities on the project really are, especially if they have served on "traditional" steering committees in the past. In the Project Pre-Check world, each stakeholder has a clearly defined accountability for a set of Decision Areas. It is not enough to attend meetings and comment, critique or criticize project performance. They now have to achieve agreement and closure from the other stakeholders. In addition, they are responsible for ensuring their agreement and commitment to Decision Areas for which the other stakeholders are accountable.

FastPath demands clear accountability and commitment. That ensures the stakeholders are engaged and on side, all relevant facets are considered and the key decisions are identified and addressed. It provides a framework for effective project launch. It enables early intervention project assessment, comprehensive planning, ongoing project control and a rational project completion process. *FastPath* is the mechanism that brings stakeholders together to achieve success on all fronts—to deliver the planned change and the expected benefits on budget, on plan and with the quality required.

Fastpath Process

The *FastPath* process provides a defined pathway for understanding and managing the impact and risks associated with a change. It provide the means for the project manager and other stakeholders to shape and contour the change over time to deliver a responsive, cost-effective, quality solution that provides optimum value to the organizations involved. You will notice a degree of commonality among the stages:

- o They are intended primarily for change agents. The change agent is the primary driver of the *FastPath* stages and needs to facilitate the effective execution of each stage as needed.
- o While the *FastPath* process presents the stages in a serial fashion, in reality each stage, or portion thereof, should be used as and when it's needed. For example, if a stakeholder departs for any reason, execute the Identify and Engage Stakeholders stages to fill the void. If one of a number of planned releases is implemented, execute the Guide Completion stage. Etcetera.
- o The stages use the same Decision Area catalogue.
- o They include the Write-in feature which allows stakeholders to add Decision Areas not covered in the base catalogue.
- o They use questionnaires to solicit information from stakeholders.
- o They consolidate the questionnaires for stakeholder presentation and review.
- o Action items from each stage are resourced, scheduled and tracked by the project team(s) using the project and change management practices that are being applied to manage the overall project.

By moving the follow-on work outside of Project Pre-Check *FastPath*, the burden on the stakeholders is reduced, allowing them to focus primarily on decision making and direction setting.

Identify Stakeholders

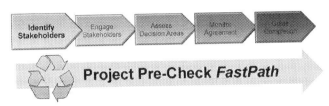

Project Pre-Check *FastPath*

"The beginning is half the whole."

Pythagoras—Greek philosopher and mathematician

Research by Standish, Gartner and others underlines the importance of getting the right people actively involved in a major business or technology change. It is the key factor for successful delivery. So, who needs to be involved and how much choice do you have, as a change agent, over their selection?

Typically, someone who wants to launch or promote a change starts thinking about who needs to be onside (the stakeholders) for the change to happen. The individual(s) participating in this deliberation may be centrally involved in the change effort or, they may have no ongoing involvement at all. For example, if you're fomenting a rebellion, chances are you'll be a central character going forward. However, if a company has dropped your favorite breakfast cereal, you'll organize resistance and publicity campaigns until the company changes its mind. After that, all you need to do is sit back and enjoy the fruits of your labour. Nevertheless, in both situations, other people need to get on board with the cause to advance the project and see it through to completion.

The Identify Stakeholders stage should be facilitated by the change agent in cooperation with the apparent sponsor. As the exercise progresses, another individual may be identified as the sponsor but at least the deliberations on the makeup of the stakeholder group will have benefited from another's input.

Fastpath Process

As a change agent, you may arrive on a project after someone else has already made the stakeholder selections. Or you may be appointed right at the start and have a clean slate in terms of stakeholder selection. In either case, you need to do your due diligence! Make sure you agree with the selections. If you don't, you'll need to identify and sell your own choices! Everything else after this point depends on getting the stakeholders right. Consider the following factors when making your recommendations:

- o Accountability—if there's someone within the organization that is already naturally identified as the change leader/owner/driver, that individual could be a rational choice for sponsor, subject to the span and capability assessments. Select the other stakeholders based on the sponsor choice.
- o Span—stakeholders should be at an appropriate level organizationally to guide a project to a successful conclusion. A manager would be an appropriate sponsor for a change that is largely constrained within his or her department. The other stakeholders would be peers or below. Changes that impact beyond the department to the division or across the enterprise need stakeholders at higher organizational levels.
- o Capability—an effective stakeholder should have the skills and capabilities to handle the change in question. Sometimes, an individual's capabilities may not be known or evident. Consider the following questions to get a better handle on a candidates capabilities:

  - Communication history
  - Clarity of goals
  - Level of dissatisfaction/need
  - Personal impact
  - Public and private influence
  - Level of understanding
  - Implementation history
  - Priorities
  - Confidence
  - Business and technical understanding
  - Resource impact
  - Relationships
  - Commitment

If you find that a stakeholder is deficient in one or more of the above areas, it's not likely that you will willingly reveal your conclusions to that individual. So what do you do with the insights you've garnered. The first step is to tailor your plan to address the deficiencies. If a stakeholder has poor communication skills or a history of not communicating as often as necessary, to the right individuals, in the appropriate format and venue, arrange support and time to address the gaps. If a stakeholder has a history of abandoning in flight projects, that's an opportunity for a face to face chat about the importance of ongoing participation. Here you can use *FastPath's* requirement for stakeholder agreement from inception through completion to foster the necessary continuing involvement. Of course, leveraging the knowledge and relationships of the other stakeholders is always an option.

Let's look at the four roles in the Project Pre-Check practice that make up the stakeholder group (sponsor, change agent, target and champion) and see what the choices are. You need at least one stakeholder in each role. If an obvious champion is not readily available, the stakeholders should consider developing the role. Typically five to ten stakeholders are required for ongoing involvement.

Sponsors

Typically, a Sponsor is the manager who launches the change and has the economic, logistical and political power to make the change happen. Sponsorship can't be delegated because the power and authority vested in an individual relative to a particular project flows from the person's position in the organization. To find the change sponsor, look for the organization, and the manager within it, most closely aligned with the planned change. If there's someone within the organization that is already naturally identified as the change leader/owner/driver, that individual could be a rational choice for sponsor, subject to span and capability considerations.

Don't shy away from placing the sponsor mantle on senior executives either. If the project has a broad impact across an organization, it will need a senior level manager to navigate the politics and overcome the natural resistance to change. In one situation I'm familiar with, a massive project was launched to address strategic, long term needs. The project affected every aspect of the company's operations including relationships with its clients and suppliers. The senior executives agreed to bring in an external resource with lots of expertise in the industry and anoint him sponsor. Understandably, the arrangement didn't work well. He had no organizational authority. He had no operational accountability. He hadn't built credibility within the organization. He had few established relationships to leverage. After nine months of wasteful expenditures and little progress, the CEO recognized the problem and appointed the COO as sponsor. Initially reluctant, the COO grew into the job with the support of the other stakeholders and guided the initiative to a successful conclusion.

So, what choices do you have as a change agent if the sponsor is simply not interested in being involved, or is making the wrong decisions, or is, in other ways, jeopardizing project success? This is one of the toughest stakeholder performance problems to deal with. Very often, sponsor performance issues aren't dealt with until after the project has failed, not a terribly useful outcome for the other stakeholders involved.

In some cases, a new sponsor can be assigned by going up or down within the same organization. For example, if the change is a new product launch and the current sponsor is the product development manager, that manager's boss could be the replacement sponsor. The other option is to move another manager into the organization to take on the functional responsibility and, as a result, pick up the sponsor mantle.

Obviously, in most cases of sponsorship change, someone has to go over the sponsor's head to make the change happen. That usually requires a very committed and cohesive stakeholder group. The upside for that effort is the potential for a new, committed and involved sponsor. The downside is the time, effort and energy devoted to sponsor replacement, usually at the expense of the project.

I know of one small organization that took on a multi-million dollar refurbishment of their administrative systems and appointed the VP, Administration as the sponsor. He was overly aggressive, abusive to anyone who opposed him and ill equipped to guide the change to a successful completion. He approached the Board of Directors on multiple occasions for incremental funding in his usual aggressive and adversarial style. The Board directed the CEO to rein him in but nevertheless granted his funding requests. The sponsor was never replaced. The project ultimately cost over four times the original budget. A strong, committed stakeholder group may have been able to nip this problem in the bud.

The following case is an example of the kind of results a super sponsor can deliver and how a PM can leverage that talent to deliver superior results.

Delivering Portfolio Management Light

The Situation

This 2000 person technology infrastructure operations group, part of an international financial services organization, had just completed a major reorganization along functional lines. Because they did not have a centralized, consistent, scalable and repeatable approach for managing their IT project portfolio, the restructuring resulted in ad hoc procedures for submitting, assessing, authorizing and monitoring change initiatives and a sizeable backlog of changes looking for approval, funding and resourcing.

The Goal

The organization set out to design and roll-out an end-to-end project portfolio process that would support the leadership team in their decision-making activities related to infrastructure investments. The developed process was to leverage industry best practices and available technology to achieve the following objectives:

- Ensure proposed initiatives were proactively socialized with all impacted stakeholder groups to solicit buy-in and support.
- Provide a consistent and repeatable approach for capturing required information for effective management decision making.
- Build and manage a project portfolio to maximize the returns from the investments in infrastructure expansion and improvement initiatives.
- Deliver and implement an operating solution within four months to address a considerable demand backlog.

The Project

The initiator and sponsor of the project portfolio venture, one of four VP's in the group, sold the project to his peers and to the senior VP who had overall responsibility for the infrastructure operations organization. He brought in a Director with whom he had previously worked to act as the first portfolio manager and be the first target. He also contracted with a consulting firm familiar with Project Pre-Check to act as the change agent for the design and implementation activities. In addition, approximately thirty directors were identified as targets of the change. In the future, they would have to understand and use the new practice to initiate projects that exceeded a certain threshold and required or impacted resources in other parts of the organization.

Fastpath Process

The sponsor selected Project Pre-Check as the source framework and specified the use of MS Office and related products for the initial launch. He chose Project Pre-Check because its Decision Framework is built upon guidance from numerous well-established and recognized industry frameworks and methodologies (e.g. ITIL, CobiT, PMBoK, SEI, ValIT, Gartner, etc.).

The sponsor worked with the targets and change agents to agree on relevant decision areas from among the 125 included in Project Pre-Check. Twenty decision areas were ultimately selected for the initiation stage and a further twenty five were selected for the planning stage submission. An Excel template was developed to capture the input. A prioritization model was built, again in Excel, to assess the multiple submissions, weight, rate and rank and generate the summary charts for the decision makers. In addition, MS Project was used to manage the resource demands using its Excel import capability and SharePoint was used to source the templates and capture the submissions.

Support materials were developed including the process models, a process guide, terms of reference and cheat sheets. The targets were engaged in a number of sessions throughout the development of the materials and provided with a half day review of the finalized process.

The Results

The first senior management portfolio review including twenty-four project submissions was conducted three and a half months after the project launch. With a few minor teaks along the way, the portfolio management light solution continues to meet the organization's needs. Not bad for a quick and dirty solution!

How a Great PM Helped

The PM on this project did four things very well:

- He involved the sponsor and gave him free reign throughout. The sponsor was an enthusiastic, respected and committed leader who knew what he wanted. The targets knew very well not to resist for very long or they would get flattened by the steamroller. The PM gave the sponsor the information he needed and stepped out of the way.
- He engaged with the targets frequently, addressed their concerns as best he could and leveraged their suggestions so that it was obvious to everyone that the solution was a collaborative effort.

- He settled for "good enough". He could have spent months evaluating and debating the selection and definition of the decision criteria. He could have spent many more months on the weighting, rating and ranking exercise. Instead, he made sure the 90% solution was in front of the stakeholders quickly and wrapped up debate expeditiously. Leveraging Project Pre-Check provided a significant advantage because it helped establish the boundaries and provided a ready, comprehensive source for decision criteria.
- He used the available technologies within their capabilities. He ensured the tools were used for what they were good at and suitable for. He didn't invest a great deal of time and money getting the technology up and running. It's easy to run and relatively easy to change and it won't cost much to get out if a decision is made to go to something more robust in the future.

Change Agents

A change agent (often, but not necessarily, a Project Manager or Project Executive) is responsible to the sponsor for implementing the change. Authority is focused on determining how to deliver according to the sponsor's mandate and the wishes of the other stakeholders. That's why executive support and user involvement are always at the top of the success factors list. It is the sponsors and targets that must make the critical calls on the 5 W's. Certainly the PM needs to provide insight and feedback on those discussions—the impact on implementation and operating costs, time, resource, quality, complexity, etc. The PM can object strenuously to a decision. He or she can even escalate if the matter isn't resolve satisfactorily. But the final decision will rest with the sponsor.

The change agent role is the only one of the stakeholder roles that has no limitations on selection. Perhaps that's why the project manager is so often the "fall guy" for project failure. The change agent can be an internal staff member, an external resource, a vendor resource, a business resource, an IT staff member or any other reasonable source of expertise. Often, projects have more than one change agent to address the breadth of the skill needs.

The best insurance a change agent can have for a successful project outcome is a committed, actively engaged stakeholder group. And, of course, if there is an effective stakeholder group in place, there's less chance a change agent, and the project, will get into trouble

Fastpath Process

Here's an example of one change agent with no prior project management experience who managed to deliver a major, global project successfully by using Project Pre-Check's building blocks.

Anybody Can Manage a Project

The Situation

A global mining company had a number of Enterprise Resource Planning (ERP) solutions in place in various regions and departments as a result of their growth by acquisition. Work was in progress to standardize their ERP platforms. As part of that standardization effort, the organization also planned to implement a global standard for segregation of duties (SOD) to meet required compliance standards (i.e. Sarbanes-Oxley).

The Goal

The company planned to develop and implement a global SOD standard that would be used to support the management of controls for regional ERP applications and to support the management of SOD risks into the future. The project would also benefit other functions including compliance.

The Project

The Finance organization launched the SOD standards initiative and developed a comprehensive SOD matrix over a period of eight months. A Working Group was established to move the project forward and included Finance department managers and staff with some consultation with the head office IT organization.

As Finance was wrapping up their SOD design, the head of Global Internal Audit heard rumblings from the regions concerning the work being done, its complexity and the regions' lack of involvement to that point. He proposed to the Finance VP that an audit be done to assess the performance of the project to date, identify gaps and target opportunities to provide the foundation for a successful implementation. His recommendation was approved.

Internal Audit launched the project audit using Project Pre-Check's Diagnostic process. The assessment involved a review of the processes used to guide the SOD work and the project deliverables to date. It also involved interviews with selected members of the Working Group and additional staff in the regions regarding their views on progress to date and thoughts and suggestions on future plans. The interviews addressed the interviewees' level of agreement on a selected subset of Project Pre-Check's Decision Areas (60 of the 125) covering the nature of the planned change, the environment within which the change would be implemented, organizational processes and practices that could be leveraged and the management of the project itself.

The interview results showed that the SOD project's overall level of stakeholder agreement on the project was 2.4 on a scale from 1 to 5, where 5 was completely in agreement with the decisions reached. The results indicated that the stakeholders interviewed were not in synch on the 60 Decision Areas addressed. Not a great recipe for success!

The results identified 43 of the 60 Decision Areas in the assessment (70%) as areas of divergence (at least one of the stakeholders was not in agreement with how a decision area was addressed) and 8 areas (17%) where a gap existed (where the majority of stakeholders expressed a lack of agreement). The responses reflected differing views on the scope of the project, costs, benefits, the target dates, the sponsor, project manager, decision making responsibilities and governance.

The audit took about four weeks to complete. The audit leader presented the findings and recommendations to the audit committee. The recommendations included:

- Confirm the sponsor
- Form a stakeholder group including comptrollers from all regions
- Appoint a project manager
- Work towards full agreement on all 60 Decision Areas included in the audit.

The Results

The sponsor was confirmed. A stakeholder group was formed and operated through to the completion of the project. A project manager was appointed to carry the project forward. She was a recent hire in the Finance organization with accounting qualifications and a financial background but no formal project management training or experience.

Fastpath Process

The project was delivered successfully in all regions according to the budgets and dates negotiated with the head office groups and with each regional comptroller. At implementation, all 60 of the Decision Areas addressed in the audit averaged 4 or greater, indicating very strong agreement among the stakeholders on the decisions made. All gaps and areas of divergence had been eliminated. The members of the stakeholder group were unanimous in their praise for the job done by the PM. Not bad for a PM with no prior project management experience!

How This Great Change Agent Helped

The senior management attention initiated by the project audit and the subsequent actions taken on the audit recommendations provided the PM with a huge starting advantage. She had a designated sponsor and she had a committed stakeholder group representing the organizations affected by the change. As well, because everyone knew she had been selected by senior management to guide the project, she was perceived to have the authority to act, to acquire the necessary staff and other resources and quickly push through the needed decisions.

To her credit, the PM used those assets effectively and took the following additional actions that were necessary to get the job done:

- She determined the sponsor's expectations regarding ongoing oversight on costs, benefits and timing and on existing corporate practices that he expected to be used. She helped him articulate his thoughts on how scope, change, issue, quality and risk management should be handled and confirmed his criteria for calling the project complete.
- She engaged the sponsor to review these expectations with the other stakeholders and gain their agreement, facilitating collaboration and compromise where needed.
- She established roles, responsibilities and operating protocols for the stakeholder group, including the sponsor, targets, the champion and her own role as change agent.
- She used the 60 Decision Areas addressed in the project audit as the basis for a report card tracking stakeholder agreement levels over time.
- She modified the standard project reporting template used by the IT organization to address the sponsor's information needs and further adapted the practices to address some unique concerns of the other stakeholders.
- She worked with the other stakeholders to develop an overall plan and approach that met the overall needs of the organization and addressed the requirements of each region.

- She went the extra mile to ensure the regional comptrollers from far off places (Asia Pacific, Australia and South America) were involved and up to speed on developments even though it meant some extra long days for her. It paid dividends in open and honest dialogue over the course of the project.

The bottom line: she was successful because she worked effectively with a committed, knowledgeable stakeholder group and she used and adapted existing frameworks and practices to guide the project to completion.

Targets

Like the sponsor role, an individual is required to fill a target role because of his or her position in the organization, directing individuals or groups who must actually change the way they operate and behave for a change to be successful. Targets include managers of departments within the initiating organization, such as marketing, sales, operations, manufacturing, information technology, finance, human resources and others. Targets can also include managers who are external to the organization initiating the change, such as customers, partners, vendors and distributors.

Targets should be selected and assessed on the same criteria as other stakeholders: accountability, span and capability. Also, never forget that a target's first responsibility is to ensure the organizations, people and functions he or she manages operate effectively and adhere to whatever commitments are in place (budget, service level, quality targets, etc.) after the change is implemented. You will find targets objecting to certain project features, functions and capabilities. That's great! That's their job. The collaborative process required to address those objections will ensure a better solution that has the full support of all stakeholders. And, if for some reason the stakeholders can't resolve a dispute, escalate the issue to a higher, common authority.

Failure of one or more targets to be fully involved and engaged in a change effort can have disastrous effects on a project yet the options available to the other stakeholders to remedy the situation are limited in the same way as sponsor replacement.

I recall one project, a new product launch, sponsored by the Marketing VP, which ran into significant challenges because of a recalcitrant target. The VP Administration, whose staff would need to develop and operate new processes to support the product, didn't agree that the product was necessary and didn't commit managers and staff to work on the initiative. The result was a stalemate that delayed the project by more than six months. The stalemate was resolved by escalating to the CEO, a strong supporter of the planned new product, who offered the reluctant target a choice—get on board or get out! The target capitulated.

Here's another example of what can happen when the target stakeholders don't rise to the occasion and provide a counter weight to as over-aggressive sponsor.

Collaboration in a Command and Control World

The Situation

In 2000, this medium sized, mid-west based manufacturing concern launched an in-house development effort to automate and enhance a core administrative function. It was a mish-mash of older technologies and manual processes that were costly, error prone and time consuming. The new application, developed in Visual Basic version 6 (VB6) back then, leveraged the in house development group's experiences with earlier versions of the language and delivered the expected improvements in quality and performance.

Unfortunately, in 2008, Microsoft ended support for VB6 and replaced it with its .NET offering. The VP Administration at the firm was aware of the impending loss of support but declined to take any action until just before vendor support for the product ended. He then challenged the CIO to find a way out of the dilemma as quickly as possible and at the lowest cost possible.

The Goal

The VP Administration, the sponsor of the undertaking, established the following objectives for the project:

- Deliver a supported solution in six months or less
- The costs should not exceed $250,000
- There should be no changes to the business user experience. They should be able to operate exactly as before.

The Project

The CIO, with the sponsor's agreement, appointed an experienced internal project manager to lead the effort. She identified and met with the other stakeholders who needed to be involved. Collectively they met with the sponsor to review his expectations and discuss some of the needs and opportunities the other stakeholders had in mind.

The sponsor restated his three objects. He rejected all other suggestions from the other stakeholders. He made it perfectly clear to everyone in the meeting that there would be no further discussion on the scope of the project. A follow-up meeting between the sponsor and project manager produced the same outcome but the sponsor added an additional requirement. He wanted an external company to bid on the project. When the project manager pointed out that that would add additional time, cost and risk, the sponsor exploded and challenged the project manager to get the job done within his expectations.

This is probably where the project manager should have removed herself from the project. However, she motored on. She worked with the other stakeholders and their staff to develop a Request for Proposal, worked with the CIO and other IT managers to identify two external candidates with solid track records and worked with those contractors to help them understand the challenges ahead. She also facilitated the development of a proposal from the internal development group. The proposals:

- External Company A—$900,000 and 10 months duration
- Internal Development Group—$600,000 and 7 months duration
- External Company B—$500,000 and 6 months duration.

The project manager, with the support of the CIO and other stakeholders, recommended the Internal Development group's proposal over the two external proposals based on a thorough, objective, fact based review. The sponsor picked External Company B's proposal. In spite of considerable debate and discussion at the executive level, the VP Administration got his way. The contract was awarded to External Company B which immediately assigned its own project manager. The internal project manager was assigned to another internal project. She was lucky!

Fastpath Process

The Results

The project was a mess! It ended up costing over $1 million and took over a year to complete. Much of the overrun was due to rework. The sponsor demanded that the contractor's project manager find ways to cut costs from their original $500K estimate. He personally cut out the costs associated with the participation of the internal development group in plan and code review and unit and integration testing. He also personally cut the involvement of the business unit staff in the same activities. He rejected any attempts to take advantage of the .NET platform and declined to consider changes to the application the business proposed to improve overall performance.

This sponsor from Hell should have looked in the mirror to place blame. Instead, he threatened the contractor with bad press and legal action. The contractor ultimately agreed to swallow most of the overrun. He also publicly criticized the other business and IT stakeholders for failing to identify competent external contractors and doing their part to ensure a successful outcome.

In this situation, there was no collaborative stakeholder group. In spite of the efforts of the internal PM to engage the other stakeholders, the sponsor's frame of reference was command and control—I say, you do! No sponsor, regardless of how experienced, informed and qualified has all the information needed to make all the right decisions necessary for managing a change. This sponsor's unwillingness to leverage the knowledge and insight of the other stakeholders was the number one cause of failure.

In this situation, the internal development group and the two contractors relied on their experiences with similar technology upgrades to provide a proposed roadmap for the work ahead. The roadmaps presented a rational series of steps and decision points, the participants in each step and the roles and responsibilities for each participant. When the sponsor started dictating what steps would not be done and what parties would not be involved, any opportunity to leverage a proven, end-to-end process went out the window. Failure cause number two!

In this situation, the sponsor's attempt to dictate all three variables—time, cost and function—caused the contractor's PM to abandon best practices that had served the organization well in the past. Their estimating practice, their approach to planning and delivering quality solutions, planning and managing risk, always considering business and technology alternatives, phasing development and staging delivery to provide early insight and added value, and others, were all sacrificed in an attempt to please an out-of-control sponsor. Cause of Failure number three!

How a Great PM Could Have Changed the Outcome

As the saying goes "Hindsight explains the injury that foresight would have prevented". We know what's needed to deliver a successful project: a committed, collaborative stakeholder group, proven processes that can be applied or adapted to the situation at hand, and use of proven best practices to cover the change landscape.

In fact, the internal PM did most things right. She engaged the other stakeholders. She escalated by bringing her CIO into the fray. She offered a number of intelligent options to the sponsor. Unfortunately, the contractor's PM didn't challenge the sponsor's edicts and didn't attempt to escalate within his own organization or within the client's organization until the damage had already been done.

Dealing with sponsor incompetence is a challenge. The sponsor is in that role because of his or her place in the organization. A replacement sponsor typically has to come from the same organization, either in a position above or below the incumbent. Obviously, a PM needs lots of assistance to make that happen. Changing a sponsor's perspective and style towards an engaged, collaborative stakeholder model can happen if the PM and other stakeholders are able to demonstrate that it's a more effective model for achieving the sponsor's goals. However, for that to work, the PM needs the other stakeholders to be actively on the sponsor's case, the message has to be delivered consistently and repeated often until the metamorphosis happens. Finally, a great PM knows when to pull the plug. You've heard the saying "Life is too short to drink bad wine." In this case the saying should be "A PM career is too short to take on bad projects."

Champions

Earlier in this book I stated that a stakeholder was a decision maker who directed an organization that initiates, is affected by or is charged with managing all or part of a change, has the authority and responsibility to set direction, establishes priorities, makes decisions, commits money and resources and is accountable for delivering the planned benefits on budget and on plan.

That's true for stakeholders in the sponsor, change agent and target roles, but not necessarily applicable to stakeholders in the champion role. A champion is a manager or senior staff member who enthusiastically supports the change and has the power, influence and respect necessary to help bring about the necessary behavioural changes in the managers and staff that are affected by the change. Champions are champions because they have the ability to influence other stakeholders, staff and other parties, to lead the way, to help others deliver and master a business or technology change with enthusiasm, commitment and passion.

For example, using a highly respected and highly successful sales person in the champion role to promote the introduction of a new product can help other sales staff overcome concerns they may have over the affect the new product will have on their clients, performance and income.

Although they are a powerful force for change, champions can be problematic to replace. Because of the respect champions are given by managers and staff who are affected by the change—the very reason they were given the champion role in the first place—removing someone from a champion role because they're not doing their jobs or delivering the wrong message can have a very negative affect on the project. It's best to work on their methods and their message. Once again, the presence of a strong stakeholder group can be the catalyst for turning a negative into a positive.

All Others

What about all those other folks who are usually involved in overseeing, supporting, contributing and commenting on a project's makeup and performance? If they are not filling one of the fours stakeholder roles and making explicit decisions on the conduct of the project, they're not stakeholders, nor are they members of the stakeholder group. If they can't logically fill one of the four stakeholder roles, they have no place in *FastPath*!

A colleague of mine recently joined a project as a contract project manager. It involved the launch of a new financial product with a boat load of web development. She jumped into the project with both feet in response to some critical target dates. And then she ran into a number of conflicts. The VP of System Development, her boss, explained that he was the project sponsor and proceeded to dictate direction, which was at odds with business expectations. The resource manager with whom she dealt to obtain project staff, issued orders regarding the functionality, look and feel of the application, again at odds with the business direction.

The project manager took a deep breath and stepped back. She asked the individuals involved who they thought the stakeholders were and what roles they were playing. She found considerable disagreement and pursued the issues until there was unanimity. As a result of her questioning, a number of changes were made to the makeup of the stakeholder group. The business VP responsible for the product launch became the sponsor. The System Development VP had no formal stakeholder role, nor did the resource manager. Two organizations affected by the project were not initially represented in the stakeholder group. Senior managers were appointed to fulfill their target roles. According to the project manager, once the stakeholder group was reformed, the difference in clarity on goals and directions, in the speed of decision making and the level of stakeholder commitment was palpable.

Stakeholders in any role can be tough to identify and even more difficult to replace. The time and effort that's usually required to manage the replacement will distract everyone from the job that matters, delivering a project successfully. It's best to assess stakeholder capability right up front and address any apparent gaps. And to keep the stakeholders engaged, make use of the Project Pre-Check processes to monitor stakeholders' level of agreement on the decision areas critical to project success.

Engage Stakeholders

"If your actions inspire others to dream more, learn more, do more and become more, you are a leader."

John Quincy Adams

Once the stakeholders and their roles are identified in the Identify Stakeholders stage, use the Engage Stakeholders stage to get them onside. It's a necessary step to ensure they understand and commit to their roles and responsibilities, to equip them with the information they need to contribute their energies, thoughts and suggestions about the planned change and manage the project's impact on their organizations.

The stage should be facilitated by the change agent with the active leadership of the sponsor. It's the sponsor's first opportunity to build the stakeholder group into an effective force for change.

The engagement can be conducted in a number of ways depending on the nature of the project and the organizations affected. It can involve:

o   A one on one or two on one session with sponsor, change agent and each or the other stakeholders
o   An orientation session for all stakeholders, preferably in one physical location
o   Any other form of communication that allows the sponsor to address the nature and rationale for the planned change and confirm the commitment of the other stakeholders to the journey.

Whatever approach is used, it should also provide the stakeholders with a forum for understanding and clarifying the planned change and their roles in ensuring a successful venture.

The engagement exercise should include the following topics:

o   The details about the planned change. Ideally, the change agent should work with the sponsor before the engagement sessions to address the Decision Areas in the Change Domain at a minimum. That will help the sponsor develop a fully formed view of the change that can be reviewed with the other stakeholders.

o   The identified stakeholders and their assigned roles and responsibilities.

o   The purpose and the five stages of the *FastPath* process.

o   The four Decision Framework Domains—Change, Environment, Assets and Project and the fourteen *FastPath* Factors included in the Domains.

o   The Decision Areas within each Factor. Stakeholders will be asked to consider 50 *FastPath* Decision Areas in the Assessment stage. Using the Decision Areas in the Change Domain is a good starting point to give a sense of the kinds of things that will be considered. If the sponsor and change agent have had an opportunity to elaborate on those Decision Areas, so much the better.

o   Sample Assessment stage questions and responses. This is the heart of the engagement session and should include the following steps:

- Review the relevance question. It asks each stakeholder to indicate, in their view, how relevant each Decision Area is to the project being assessed. The possible responses are:

    — Not Relevant—the Decision Area in question provides no value to the project and is not impacted by the project in any way.
    — Relevant—the Decision Area provides value to the project or is impacted by the project and should be included.
    — Don't Know Yet—it's unclear whether the Decision Area provides value to or is impacted by the project.

- Review the Accountability question. The stakeholder assigned accountability is accountable for overseeing subsequent effort on impact assessment, the identification and review of alternatives and development of recommendations for review by the other stakeholders.

Fastpath Process

o The table below suggests an initial assignment of accountabilities by Decision Framework Domain, Factor and stakeholder role.

Domain	FastPath Factors	Stakeholder Role			
		Sponsor	Change Agent	Target	Champion
Change	Dimensions	√			
	Stakeholders	√			
	Investment Evaluation	√			
	Quality	√			
Environment	Future State			√	
	Business Plan			√	
Assets	Resources		√		
	Project Management		√		
	Security Administration		√		
	Technology Operations		√		
Project	Planning		√		
	Organization		√		
	Control		√		
	Communication		√		

Table 2—Stakeholder Accountabilities by Domain & Factor

- Review the Stakeholder Agreement question. This is the core of Project Pre-Check and *FastPath*! Stakeholders who agree on the key decisions will be successful. By managing the level of agreement on each Decision Area from inception through completion, issues will be identified early and resolved expeditiously, either through collaboration or escalation. In addition, the Agreement question provides stakeholders with a *raison d'être* for ongoing project involvement. Possible responses are:

1. Don't know or disagree
2. Mostly disagree
3. Agree somewhat
4. Mostly agree
5. In complete agreement

The responses need to be based on a thorough understanding of available documentation and/or empirical evidence to ensure all views are based on the same scale. Having a stakeholder select a *disagree* response without being able to point to a specific document, slide deck, presentation or discussion doesn't cut it. After all, if there is a low overall score for a given Decision Area or a divergence of responses, the reasons behind those responses need to be explored, with appropriate reference to the source materials.

- Review the Write-in option. Write-ins can be added by each stakeholder to record decisions pertinent to the change that have been or need to be addressed and are not included in the base covered in the **FastPath** Decision Framework. That can include Decision Areas covered in the full Project Pre-Check Decision Framework but not included in **FastPath**. It can also include Decision Areas that are unique to an organization, industry, region or country.

o Agree on the target date for the completion of the Assessment questionnaires. The total time to complete the questionnaire usually does not require more than an hour per stakeholder.

Assess Decision Areas

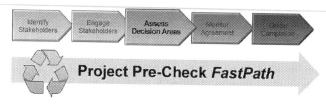

"Truth is the most valuable thing we have."

Mark Twain

The objective of the Assessment stage is to ensure that project stakeholders are in agreement on the decisions that need to be made and the decisions taken. That agreement is achieved by determining each stakeholder's views on Decision Area relevance, accountability and level of agreement, consolidating the results to identify gaps and areas of disagreement and initiating actions to address the issues identified.

The *FastPath* Assessment questionnaire is the enabler in this process. The time frame to complete the individual questionnaires is usually not more than an hour. Total turnaround time of a couple of days should be more than sufficient, in most cases, to get the completed submissions from all stakeholders.

The responses should reflect actual personal involvement, knowledge and comfort and be based on existing, available documentation or empirical evidence to reflect specific, documented positions or demonstrated capability. To get things started, the change agent should complete the questionnaire to get a sense for what the responses might be. The first time around, I'd recommend a one on one interview between the change agent and each stakeholder. Start with the sponsor if possible. That way, the change agent can address any questions and ensure consistent, fact-based responses. Most people get the exercise up front but there will always be a few exceptions. I once did a Decision Area assessment with a stakeholder who interpreted any response below a "5—In Complete Agreement" as a sign of failure on her part. After the tenth time asking her to explain what the assessment was based on and leading her to select a more appropriate response, she finally clued in.

Another common challenge deals with the number of Decision Areas. Stakeholders often ask why they are being held accountable for reviewing, understanding and ultimately agreeing with "project stuff". That can also be an issue with the change agent, who may view stakeholder involvement in "project stuff" as an intrusion on the PM's authority. The answer to both complaints is quite simple. Every one of the 50 *FastPath* Decision Areas (and all of the 125 Decision Areas in the Project Pre-Check Decision Framework) can have a specific, direct impact on project success. Therefore, all stakeholders need to be involved and in agreement on the decisions made.

Take the Cost Estimate Decision Area in the Project Domain for example. I've had a number of business stakeholders object to being involved in any way with understanding and agreeing to the estimates and underlying assumptions. That's the PM's job, they insist. Certainly they'll object when the estimate is more than they want to pay. They'll scream when the estimate is exceeded. But that's the PM's responsibility. Well, did you know that one study found almost half of all project failures examined were due to bad estimates?

Accurate estimates are a vital component of effective change management. When all stakeholders are involved and in agreement, the numbers are more comprehensive, more complete and more accurate. If one or more stakeholders aren't on side with the initial estimates or the underlying assumptions and approach, their lack of agreement generates an action item which triggers project activity to resolve the concern. The bottom line: better understanding, better decision making, a more successful project.

Consolidation of the results for review with stakeholders serves as the catalyst to get stakeholders focusing on and resolving the issues revealed. The stakeholder group should be aiming to have complete unanimity on all questions of relevance and accountability and an average agreement score of 4 or greater with no one below the 4 level as soon as possible. Obviously, at the start of a project, there will be a variety of views on relevance and accountability and bottom level agreement scores. Even if one stakeholder is overly enthusiastic about progress, the other stakeholders' responses will usually provide a counterbalance and reflect a more realistic view of the project state.

The Assess Decision Area stage should include the following steps:

Fastpath Process

o   Distribute the questionnaires to stakeholders. The questionnaire can be distributed as a spread sheet via email, accessed through a web site as either a download or editable questionnaire, handed out as hard copy, etc. The automated approaches are obviously preferred because of the ease of consolidation. As mentioned, the questionnaires can also be completed by one on one interviews to help stakeholders navigate the list of Decision Areas the first time.

Each stakeholder should complete and return the questionnaire:

- Indicate how relevant each Decision Area is to the project being assessed. The possible responses are:

    — Not Relevant—the Decision Area in question provides no value to the project and is not impacted by the project in any way.
    — Relevant—the Decision Area provides value to the project or is impacted by the project and should be included.
    — Don't Know Yet—it's unclear whether the Decision Area provides value to or is impacted by the project.

- Select one accountable stakeholder for each Decision Area. All relevant Decision Areas need to be assigned and only project stakeholders can be assigned accountability.

- Record the level of agreement on the decisions reached to date on each relevant Decision Area using the following scale:

    1. Don't know or disagree
    2. Mostly disagree
    3. Agree somewhat
    4. Mostly agree
    5. In complete agreement

The same response—1—is used whether the stakeholder doesn't know the status of decisions made or disagrees with those decisions. That's because the remedy, and a "1" response always requires a remedy, needs follow-on effort to resolve the rating.

- Add Write-ins—identify Decision Areas that you believe are affected by the project or can contribute to the project but are not included in the basic list. Record the Decision Area, a brief description and an appropriate response from 1 to 5. The Write-ins will be consolidated and presented to all stakeholders as part of the consolidated results.

o Consolidate the responses from returned questionnaires including any Write-in Decision Areas to identify the following:

- For accountabilities, total the number of selections for each stakeholder who has been identified as accountable, by Decision Area. Include Write-ins.

- For relevance, total the number of selections for each relevance category, by Decision Area.

- For agreement level, average the responses by Decision Area for all stakeholders

- Highlight strengths, gaps and areas of divergence to help the stakeholders focus on the Decision Areas that will need follow-on discussion and action:

    — Strengths—Decision Areas where the lowest response is 4 or greater.
    — Areas of Divergence—Decision Areas where the lowest response is equal to or less than 3 and the average is equal to or greater than 3.
    — Gaps—Decision Areas where the average response is below 3.

o Distribute the consolidated responses, the *FastPath* Scorecard, to all stakeholders. The Scorecard is the vehicle that is used throughout the project to monitor stakeholder positions on relevance, accountability and agreement levels and guide actions to resolve differences.

o Identify the actions required to address relevance, accountability and agreement issues as well as those responsible for each action.

Fastpath Process

Often, especially early in the project, additional investigation and discussion will be needed to address issues revealed by the assessment. For example, if the Worth Decision Area is an issue (the relevance rating is not known or the agreement level is below 3), the stakeholder accountable—usually the sponsor—will need to ensure the development of a worth position identifying how much the organization can afford to spend and review this information with the other stakeholders.

Where there are differences of opinion on relevance, accountability and agreement level, discussion among the stakeholders can often identify simple remedies. For example, where one or more stakeholders are not aware of documents and/or discussions that have taken place that support a certain position, provide them with the appropriate information and let them reassess their position.

The real fun begins when stakeholders have read and digested available information on a Decision Area, have discussed it with their staff, their peers, their bosses and have come to different conclusions about the suitability of the current recommendation.

I recall one situation where a stakeholder objected to the introduction of agile techniques on a project which already had time and complexity challenges. The issue was debated within the stakeholder group without resolution, finally escalated and the objecting stakeholder was overruled. However, he kept registering a "1—Disagree" on the Software Delivery Decision Area (a Project Pre-Check Decision Area not included in the base *FastPath* Decision Framework) looking for signs of positive contribution as the project progressed. The project began running into difficulty with the looming target dates and the project manager finally recommended that the team revert to their traditional development practice to focus all their energies on delivering on schedule. The agile practice had become a distraction. After discussion and debate among the stakeholders, they backed the PM's recommendation. The original objecting stakeholder changed his agreement level to "5—In complete agreement".

o Integrate the action items into the project plan. In Part II I mentioned that follow-on work required to address assessment issues should be managed through whatever mechanisms are being used to plan, schedule, resource and control project activities and deliverables. That way, effort required to address Decision Area concerns will be tackled in an appropriate timeframe with the right resources and monitored accordingly.

For example, if the stakeholders are not happy with the Benefits Decision Area (e.g. an agreement level less than 3), the action item should be assessed in terms the players that need to be involved (sponsor, targets, business analysts, other business and technical staff, vendors, etc.), when the activity needs to be completed (priorities, dependencies, capacities, etc.) and how long it will take. That information will be reflected in the plan as a normal project activity. The activity can be closed as complete when the stakeholder agreement level is 4 or more.

Let me stress that the members of the stakeholder group need to have absolute authority over their areas of responsibility to collaborate on the necessary decisions to see the project through to a successful conclusion. Part of the stakeholder role is to manage upward where necessary to ensure bosses are on side and in agreement on any decision that might be contentious.

Here's an example of what can happen when one's superior isn't kept in the loop. The project in question had passed through a number of control gates over a six month period, which included incremental approval of a proposed technology architecture and the target technology implementation plan. The Director responsible for the line of business architecture participated in the review and approved the relevant architectural content at each gate. After the last gate review, just prior to acquiring and implementing the recommended technologies, the VP responsible for enterprise architecture and the Director's boss got wind of the plans and pulled the plug. That change cost hundreds of thousands of dollars in rework and delayed the project by over six months.

When a stakeholder expresses his or her position on relevance, accountability and agreement level, it needs to be a commitment that represents the backing of the organization and people represented. A position can and does change, but it should never be over-ridden.

Table 3 below shows a sample *FastPath* Decision Area questionnaire. It shows the three questions that each stakeholder needs to consider and sample responses. In my experience, the vast majority of stakeholders love the process. They feel more informed, more engaged and more in control. It gives them a voice.

Table 4, the *FastPath* scorecard, shows a sample consolidation of the stakeholder questionnaires. It clearly reveals the gaps, areas of divergence and strengths in the stakeholders' views of relevance accountability and agreement level.

Project Pre-Check FastPath Assessment Questionnaire

Stakeholder Name: Sample Stakeholder   Project Name: Sample Project

Domain	Factor	Decision Area	Description	Decision Area Relevance — Mark an "X" in the appropriate — Not relevant	Relevant	Don't know	Accountability — Mark an "X" below the Stakeholder most accountable for each Decision Area — Stakeholder One	Stakeholder Two	Stakeholder Three	Stakeholder Four	Stakeholder Five	Stakeholder Six	Stakeholder Agreement — 1 - Don't know or disagree / 2 - Mostly disagree / 3 - Agree Somewhat / 4 - Mostly Agree / 5 - In Complete Agreement
Change	Dimensions	Burning Platform	current conditions that necessitate a change		X		X						5
Change	Dimensions	Opportunity	business opportunity or need		X		X						5
Change	Dimensions	Goals	business goals & objectives		X		X						2
Change	Dimensions	Worth	business worth - how much organization is willing to spend		X			X					2
Change	Dimensions	Requirements	required functions, features & capabilities		X			X					3
Change	Dimensions	Benefits	planned benefits - tangible & intangible		X								2
•	•	•	•	•	•	•	•	•	•	•	•	•	•
Project	Control	Change Tracking & Reporting	actual change activity and impact		X					X			4
Project	Control	Issue Tracking & Reporting	actual issue activity and impact		X					X			5
Project	Control	Project Completion	factors and results that will constitute completion		X					X			3
Project	Communication	Monitor Effectiveness	monitor updates on progress & feedback from all target audiences			X				X			1
Write-ins													
Assets	Business Operations	Chart of Accounts	list of accounts used to capture and track financial information		X						X		1

Table 3—Sample FastPath Questionnaire

Project Pre-Check FastPath Scorecard

Project Name: Sample Project

Relevance — Number of selections by Relevance category

Accountability — Number of selections for the Stakeholder most accountable for Decision Area

Stakeholder Agreement
1 - Don't know or disagree
2 - Mostly disagree
3 - Agree Somewhat
4 - Mostly Agree
5 - In Complete Agreement

Domain	Factor	Decision Area	Description	Not relevant	Relevant	Don't know	Acc S1	Acc S2	Acc S3	Acc S4	Acc S5	Acc S6	Agr S1	Agr S2	Agr S3	Agr S4	Agr S5	Agr S6	Average	Action Required
Change	Dimensions	Burning Platform	current conditions that necessitate a change		3	3	6						5	3	5	2	2	2	3.2	x
Change	Dimensions	Opportunity	business opportunity or need		3	3	6						5	3	5	2	2	2	3.2	x
Change	Dimensions	Goals	business goals & objectives		6		6						2	3	5	4	3	5	3.7	x
Change	Dimensions	Worth	business worth – how much organization is willing to spend		6		6						2	3	3	3	4	3	3.0	x
Change	Dimensions	Requirements	required functions, features & capabilities		6		6						3	2	3	2	2	5	2.8	x
Change	Dimensions	Benefits	planned benefits – tangible & intangible		6		6						2	2	3	2	2	3	2.3	x
⋮	⋮																			
Project	Control	Change Tracking & Reporting	actual change activity and impact		6					6			4	2	5	4	3	4	3.7	x
Project	Control	Issue Tracking & Reporting	actual issue activity and impact		4	2				6			5	1	3	1	3	3	2.7	x
People	Control	Project Completion	factors and results that will constitute completion		3	3						6	3	2	2	1	3	3	2.3	x
Project	Communication	Monitor Effectiveness	monitor updates on progress & feedback from all target audiences		6							6	1	2	2	1	2	2	1.7	x

Write-Ins

Domain	Factor	Decision Area	Description	Not relevant	Relevant	Don't know	Acc S1	Acc S2	Acc S3	Acc S4	Acc S5	Acc S6	Agr S1	Agr S2	Agr S3	Agr S4	Agr S5	Agr S6	Average	Action Required
Assets	Business Operations	Chart of Accounts	list of accounts used to capture and track financial information		4	2						6	1	1	1	1	1	1	1.0	x

Legend: Highlighted Areas Require Action

Table 4—Sample FastPath Assessment Consolidation

Monitor Agreement

Project Pre-Check *FastPath*

"Kites rise highest against the wind—not with it."

Sir Winston Churchill

The objective of the Monitor Agreement stage is to guide the multitude of variables involved in major business and technology change to deliver expected results. It is a juggling act that continues from the beginning to the end of the project and requires the active involvement of stakeholders to make the decisions necessary to ensure the goals of the change are realized. It is the stage where stakeholders balance the competing demands for functionality and capability against the need to deliver a timely, cost-effective, quality solution within budget and benefit targets.

The Monitor Agreement stage mandates that every submission reviewed, every change considered and every issue raised is addressed in the context of one or more Decision Areas. For example, if there's an issue with acquiring a certain skill set for the project, the first question should be "what Decision Area(s) are involved?" The answer: consider it in the context of the Resource Plan Decision Area in the Project Domain. The Decision Area is obviously relevant. The accountability is presumably with the change agent. If the change agent's plan to solve the issue doesn't satisfy some or all of the other stakeholders, they'll record an agreement level below 4 which will generate an action item to resolve the issue. That enables stakeholders to revisit and reassess relevance, accountability and agreement levels as needed over the life of the project to confirm that it's still heading in the right direction. It also ensures that the ***FastPath*** scorecard provides a comprehensive and ongoing record of stakeholders' decisions and project progress.

The primary focus of the Monitor Agreement stage is the review and acceptance of the Decision Area specifications, produced by stakeholder group and project team members and others, which relate to the *FastPath* Decision Areas. A Decision Area specification is a document or set of documents and other deliverables that specify, in a complete, precise, verifiable manner, the new or additional requirements, design, behavior, or other characteristics of the elements underlying a Decision Area. That includes systems, components, products, processes, practices, results, or services and, often, the procedures for determining whether these provisions have been satisfied. The specification documents will form a vital part of the project documentation throughout the project life cycle. The content may be altered at various stages and should be managed accordingly to keep it ever green.

The Monitor Agreement stage should include the following steps:

o Review the Decision Area specifications. Each stakeholder needs to understand the specifications being submitted for review and confirm in his or her own mind the degree to which the specifications satisfy the needs, plans and goals of the project from their frame of reference (i.e. the organizations and people whose interests they represent).

Does every stakeholder need to review every document in detail, make notes on points of agreement, issues and concerns and seek individual clarification where necessary? Of course not! In many cases, stakeholders will rely on others to provide insight and comments on the specifications. However, the following four conditions should be satisfied for a stakeholder to arrive at a level of agreement:

- Documentation and/or empirical evidence must exist to support a specification. Deliverables such as prototypes, models, diagrams, videos can all be suitable specification forms for consideration by stakeholders, as long as they address the need.
- The specification has been reviewed by someone with sufficient knowledge of the Decision Area from the stakeholder's frame of reference to arrive at an informed assessment.
- The stakeholder has reviewed the documentation personally or has reviewed the informed assessment provided by a knowledgeable and trusted source.
- The stakeholder is personally committed to and can defend his or her stated level of agreement.

Stakeholders can delegate specification review but they can't delegate the decision. If they do, they give control and direction of the change to others and abrogate their responsibilities as stakeholders.

o   Determine the level of stakeholder agreement. Stakeholders need to reach agreement that the specifications are sufficiently complete and appropriate for the planned change. The level of agreement is expressed using the same scale used in the Assess Decision Areas stage.

Stakeholders should consider the association with other Decision Areas when reviewing a Decision Area specification. For example:

- Is it consistent with the project's goals, objectives and quality targets?
- Is it within the project's current scope, target dates, resources and budget?
- Does it positively or negatively affect worth and benefits?
- Does it have any positive or negative impact on the project's risk profile?
- Does it change previously established Decision Areas relevance ratings?
- Does it require the involvement of new or additional parties?

Essentially, the above questions relate to other Decision Areas. And, that's the challenge and the value of specification review—to understand the inter-relationships and inter-dependencies among the Decision Areas.

For example, the specifications for the Technology Alternatives Decision Area in the Project Domain may have looked at a number of technology options and recommended the acquisition of a software package or service through a third party provider. That will have an impact on the External Relationships Decision Area in the Environment Domain, requiring contract negotiations and contract management specifications to be developed or amended.

A specification that impacts any other Decision Areas should receive an agreement rating appropriate to the impact. If the affect is negative, the agreement level would be 3 or lower. The low rating will initiate further exploration of alternatives to mitigate the impact or lead to the submission of a change request to alter the basic project parameters. If the affect is positive, the rating would be 4 or higher but may require a change in the agreement level of the affected Decision Area(s).

Each stakeholder's level of agreement can be submitted individually via email, phone or some other mechanism for consolidation and subsequent review. The opinions can also be presented and recorded during the course of a meeting.

o   Review the progress reports from the project team(s) and look for any indicators that may affect the change fundamentals—function (including quality, capability and benefit targets), time and cost.

- Identify possible repercussions on previously established Decision Area relevancy, accountability or agreement:

    — Do Decision Areas previously considered not relevant now become relevant?
    — Do Decision Areas previously assessed as relevant, now require change?
    — Do Decision Area specification agreements now require reconsideration or change?

- Reach agreement on a course of action for each instance of potential impact:

    — If Decision Area relevance is to be revisited, follow the process outlined in the Assess Decision Areas stage.
    — Ensure the previously established Decision Area accountability is still appropriate or revise accordingly.
    — If a Decision Area agreement level needs to be reconsidered, follow the process outlined in the Assess Decision Areas stage.

o   Review the affect of other influences on Decision Area relevance, accountability and agreement levels. Things change! Markets change. Competitors change. Legislation changes. People change. Technology changes. We get new insights and ideas. Progress doesn't always match plan. The Monitor Agreement stage is the stakeholders' instrument for managing those changes over the course of a project. Also, it is a primary interface point to other development, technology change, management of change and project management practices being used in conjunction with *FastPath* to manage a project.

Fastpath Process

Every change request should identify Decision Area impacts as part of the submission and approval process. Project activity, including issue and risk management, may require reconsideration of Decision Area relevance, accountability and agreement levels. As well, stakeholder issues (departures, organizational changes, lack of commitment) and external events (competitor actions, political or legislative changes, changes in organizational priorities) can change the suitability of prior decisions.

o   Update the *FastPath* scorecard to reflect the status of each Decision Area specification reviewed and affected and the impact of other influences. That includes any changes in Decision Area relevance, accountability and agreement levels. The scorecard is the primary means of communicating stakeholder decisions to the project team(s) charged with implementing the change.

To help manage the status of each element within a Decision Area, additional items can be added to the *FastPath* scorecard under the relevant Decision Areas, with appropriate accountability, relevance and agreement information.

In the example above involving technology alternatives and the affect of the recommended approach on external relationships, sub Decision Areas can be added to track and control contract negotiations (legal, operational and other aspects) and contract management (compliance, invoice handling, etc.). Or, additional Decision Areas can be accessed from the full Project Pre-Check Framework or added as Write-ins.

The Monitor Agreement stage is applied throughout the course of the project as long as there are Decision Area relevancy, accountability or agreement level issues to be dealt with. Table 5 below shows a sample *FastPath* scorecard. Notice how readily identifiable the gaps and areas of divergence are. Every action item shown on the scorecard should have a corresponding activity in the overall project plan. As the work to resolve the identified issues is completed and submitted to the stakeholder group for consideration, the stakeholders will update their relevance, accountability and agreement level assignments and those new views will be reflected in the scorecard to enable active, ongoing oversight.

Project Pre-Check FastPath Scorecard

Project Name: Sample Project

Stakeholder Agreement legend:
1 - Don't know or disagree
2 - Mostly disagree
3 - Agree Somewhat
4 - Mostly Agree
5 - In Complete Agreement

Domain	Factor	Decision Area	Description	Relevance: Not relevant	Relevant	Don't know	Acct: S1	Acct: S2	Acct: S3	Acct: S4	Acct: S5	Acct: S6	Agr: S1	Agr: S2	Agr: S3	Agr: S4	Agr: S5	Agr: S6	Average	Action Required
Change	Dimensions	Burning Platform	current conditions that necessitate a change		6		6						5	3	5	3	4	4	4.0	x
Change	Dimensions	Opportunity	business opportunity or need		4	2	6						5	3	5	2	2	2	3.2	x
Change	Dimensions	Goals	business goals & objectives		6		6						4	4	5	4	4	5	4.3	
Change	Dimensions	Worth	business worth – how much organization is willing to spend		6		6						5	5	4	5	5	4	4.7	
Change	Dimensions	Requirements	required functions, features & capabilities		6		6						4	4	5	5	4	5	4.5	
Change	Dimensions	Benefits	planned benefits - tangible & intangible		2	4	6						2	2	3	2	2	3	2.3	x
Project	Control	Change Tracking & Reporting	actual change activity and impact		6					6			4	5	5	4	4	4	4.3	
Project	Control	Issue Tracking & Reporting	actual issue activity and impact		4	2				6			5	1	3	1	3	3	2.7	x
Project	Control	Project Completion	factors and results that will constitute completion		3	3						6	3	2	2	1	3	3	2.3	x
Project	Communication	Monitor Effectiveness	monitor updates on progress & feedback from all target audiences		3	3						6	1	2	2	1	2	2	1.7	x
Write-ins																			
Assets	Business Operations	Chart of Accounts	list of accounts used to capture and track financial information		1	5						6	1	2	1	1	1	1	1.2	x

Legend: Highlighted Areas Require Action

Table 5—Sample FastPath Scorecard

Guide Completion

Project Pre-Check *FastPath*

Ehrlich's Rule: The first rule of intelligent tinkering is to save all the parts.

The objectives of the Completion stage include:

o   To establish stakeholder perspectives on the degree of success.
o   To agree on follow-up action to maximize the value of the change.
o   To identify opportunities to improve the ***FastPath*** process & Decision Areas and recommend changes to internal practices for future value.

The Completion stage is carried out as the planned change approaches its end, anywhere from 75% to 100% complete. It can also be done on a phase by phase or release by release basis. The stage involves a review of the relevant Decision Areas plus Write-ins to assess the degree to which expectations were or are being met. It does not replace the need for a normal post implementation review usually conducted with the project team(s).

The review can be completed through remote solicitation (via email for example), in a workshop or in one on one interviews.

"In the middle of difficulty lies opportunity."

Albert Einstein

The Guide Completion stage should include the following steps:

o If stakeholders agree that the Completion assessment should proceed, distribute and complete the Completion questionnaire and return for consolidation of the results.

The Completion Review uses the following response values for the included Decision Areas:

1—Did not meet goals and expectations or don't know
2—Generally met goals and expectations
3—Completely met goals and expectations
4—Exceeded goals and expectations
5—Far exceeded goals and expectations

The guidelines from the Assess Decision Areas stage also apply here. The responses should reflect actual personal involvement, knowledge and comfort and be based on existing, available documentation or empirical evidence to reflect specific, documented positions or demonstrated capability. After all, the goal is to achieve stakeholder agreement on the degree to which expectations were met. Being able to refer to something concrete is a prerequisite for reaching that consensus. A sample Completion questionnaire is shown in Table 6.

o Consolidate the individual Completion questionnaires and distribute the consolidation results. A sample consolidation is shown in Table 7. It quickly illustrates overall stakeholder sentiment, individual and average views by stakeholder and average results for each Decision Area.

o Review the consolidation results and agree on courses of action. The findings from the review of the consolidated questionnaire results can be used to initiate a variety of follow-up activity including:

- Research into one or more of the key Decision Areas where results did not match expectations to determine causes and possible remedies.
- New or additional project activity to deliver or enhance benefits, reduce ongoing expenses or improve performance or quality factors.
- Changes to the structure and application of *FastPath* to improve value.

- Changes to the structure and application of other internal processes and practices to improve the organization's ability to deliver responsive, cost-effective, quality solutions, on budget and plan consistently, over time.

  o Finally, when stakeholders are in agreement across the board, expectations have been met or exceeded and everything that can be done to maximize the value delivered has been done, it's time to schedule the celebration. Congratulations!

Project Pre-Check FastPath Completion Questionnaire

Stakeholder Name: Sample Stakeholder

Project Name: Sample Project

Domain	Factor	Decision Area	Description	Stakeholder Agreement 1 - Did not meet goals & expectations or don't know 2 - Generally met goals & expectations 3 - Completely met goals & expectations 4 - Exceeded goals & expectations 5 - Far exceeded goals & expectations
Change	Dimensions	Burning Platform	current conditions that necessitate a change	5
Change	Dimensions	Opportunity	business opportunity or need	5
Change	Dimensions	Goals	business goals & objectives	2
Change	Dimensions	Worth	business worth - how much organization is willing to spend	2
Change	Dimensions	Requirements	required functions, features & capabilities	3
Change	Dimensions	Benefits	planned benefits - tangible & intangible	2
•	•	•	•	•
•	•	•	•	•
•	•	•	•	•
Project	Control	Change Tracking & Reporting	actual change activity and impact	4
Project	Control	Issue Tracking & Reporting	actual issue activity and impact	5
Project	Control	Project Completion	factors and results that will constitute completion	3
Project	Communication	Monitor Effectiveness	monitor updates on progress & feedback from all target audiences	1

Write-ins

Assets	Business Operations	Chart of Accounts	list of accounts used to capture and track financial information	1

Table 6—Sample FastPath Completion Questionnaire

Fastpath Process

Project Pre-Check FastPath Completion Consolidation

Project Name: Sample Project

Stakeholder Agreement
1 - Did not meet goals & expectations or don't know
2 - Generally met goals & expectations
3 - Completely met goals & expectations
4 - Exceeded goals & expectations
5 - Far exceeded goals & expectations

Domain	Factor	Decision Area	Description	Stakeholder One	Stakeholder Two	Stakeholder Three	Stakeholder Four	Stakeholder Five	Stakeholder Six	Average	Action Required
Change	Dimensions	Burning Platform	current conditions that necessitate a change	5	3	5	3	4	4	4.0	
Change	Dimensions	Opportunity	business opportunity or need	5	3	5	3	4	2	3.2	x
Change	Dimensions	Goals	business goals & objectives	2	4	5	2	4	5	4.3	x
Change	Dimensions	Worth	business worth - how much organization is willing to spend	2	5	4	5	4	4	4.7	x
Change	Dimensions	Requirements	required functions, features & capabilities	3	4	5	5	4	5	4.5	
Change	Dimensions	Benefits	planned benefits - tangible & intangible	2	2	3	2	2	3	2.3	x
...	...	...	...	...	...	...	...	...	...	...	...
Project	Control	Change Tracking & Reporting	actual change activity and impact	4	5	5	4	4	4	4.3	
Project	Control	Issue Tracking & Reporting	actual issue activity and impact	5	1	3	1	3	3	2.7	x
Project	Control	Project Completion	factors and results that will constitute completion	3	2	2	1	3	3	2.3	x
Project	Communication	Monitor Effectiveness	monitor updates on progress & feedback from all target audiences	1	2	2	1	2	2	1.7	x

Write-ins

Domain	Factor	Decision Area	Description	Stakeholder One	Stakeholder Two	Stakeholder Three	Stakeholder Four	Stakeholder Five	Stakeholder Six	Average	Action Required
Assets	Business Operations	Chart of Accounts	list of accounts used to capture and track financial information	1	2	1	1	1	1	1.2	x

Legend:

Highlighted Areas Require Action

Table 7—Sample FastPath Consolidated Completion Results

Managing the Process

"It is not the strongest of the species that survive, nor the most intelligent, but the one most responsive to change."

Charles Darwin

The Project Pre-Check *FastPath* process and stages will quickly become familiar to the project manager and the other stakeholders as a project progresses. The PM is usually the one accountable for ensuring effective *FastPath* use. However, the initial learning curve, though small, can be expedited by appointing a *FastPath* facilitator to work with the PM and handle the planning, scheduling and control tasks. The chosen individual can take responsibility for arranging the sessions and managing deliverables and can leverage the experience gained to apply the practice on additional projects. Remember though, the individual selected is acting on behalf of the stakeholders and needs their full support to succeed. The selected leader can be one of the project stakeholders or someone outside of the stakeholder group, depending on the group's preference.

Using with Other Standard Methods

FastPath operates independently from other standard practices including project management, software development, technology change and management of change methodologies. It focuses on the stakeholders and the decisions they need to make to ensure the planned change is a success. Most other practices focus on particular groups outside of the stakeholder realm (e.g. project managers, software developers, etc.) Using *FastPath* ensures a cohesive stakeholder group operating from the same playbook throughout the change.

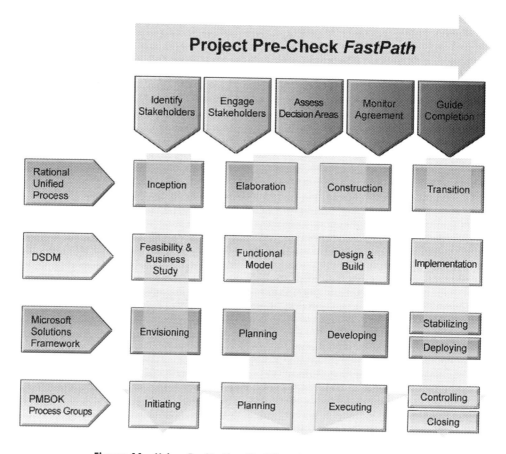

Figure 11—Using *FastPath* with Other Standard Methods

By operating independently and interfacing as appropriate, *FastPath* allows the other practices to worry less about decision making matters and to concentrate more on progress and delivery.

The diagram above illustrates the relationships among the *FastPath* stages and four other common industry practices.

The Identify and Engage Stakeholders stages creates a strong stakeholder team to guide the change through to successful completion and enables them to establish and manage the scope of the planned change relative to 50 plus Decision Areas over the entire project duration.

The Assess Decision Areas and Monitor Agreement stages provide a comprehensive view over time on the level of stakeholder agreement and identify strengths as well as gaps and areas of divergence that need immediate attention to ensure the success of the planned change. Both stages start early and continue throughout the project.

Finally, the Guide Completion stage provides a mechanism for the stakeholders to ensure they maximize the return as a project nears its end, in terms of project specific plans and expectations as well as ongoing organizational value from potential future opportunities and accessible lessons learned.

Remember too that each stage can and should be iterated as the situation warrants. If a stakeholder departs midway through a project, exercise the Identify and Engage Stakeholders stages to select the replacement and bring them up to speed. Also, the Guide Completion stage should be used on any subprojects or releases to manage value stream delivery.

Also, don't forget the actual planning and execution of the activities required to implement the decisions made by the stakeholders on the relevant Decision Areas should be included in the plans developed and managed through the other project or company practices. For example, if the stakeholders agree on a specific solution performance requirement, or a usability target, or changes to corporate reporting practices, or embellishment of the estimating practices, that work needs to be added to the project plan and funded and resourced accordingly. Decisions

Fastpath Process

may also involve the need for new stakeholders or changes to the makeup of the stakeholder group itself.

"If we want things to stay as they are, things will have to change."

Giuseppe Tomasi di lampedusa—
one of Sicily's most celebrated novelists

Leaving a Legacy of Lessons Learned

This is the perfect time and place to consider what worked and what didn't work from a stakeholder perspective and record the findings for posterity. However, the road to leaving a legacy of lessons learned is filled with pot holes and pit falls.

Project completion reviews or post-mortems are often seen as an essential best practice that enables organizations to learn from their experiences. However, all too frequently, the insights gained from post implementation reviews are lost to posterity.

There are a ton of books, periodicals and articles that address project management, software engineering and management of change disciplines and practices. But the rate of success for major business and technology changes is still well below 50%? Why? The bottom line—we lack a common mechanism where findings can be stored, examples cited and recommendations for future undertakings recorded and accessed. And, we lack the structure and discipline to ensure that our knowledge is managed and referenced consistently and rigorously.

According to Nancy Dixon in her *conversation matters* blog, "NASA learned its lesson about losing knowledge early in 1990. They experienced the sad recognition that much of the knowledge about how to build the Saturn V rocket that took the astronauts to the moon, had retired along with the engineers who had been encouraged to take early retirement". In response, NASA created the NASA Engineering Network a knowledge network to promote learning and sharing among NASA's engineers. The Network includes the NASA Lessons Learned data base, "the official, reviewed learned lessons from NASA program and projects".

Most stakeholders involved in a change aren't aware of all the best practice information out there and aren't inclined to spend the time and money to find out. They're business people, financial types, actuaries, engineers, marketing folks, business analysts, IT practitioners. They're not project management or management of change experts. They don't really understand the role they need to play and the information they need to ensure success! They just want to get the job done.

Fastpath Process

So, what do you do as a sponsor, stakeholder, PM, BA or other interested party to leave a legacy of lessons learned? There are five critical steps:

o   Identify and confirm accountability for managing lessons learned

Somebody needs to own this practice, to establish the goals and objectives, to measure performance, communicate to stakeholders and establish and drive initiatives to increase organizational value. Even open source software groups, which counts on thousands of interested volunteers to deliver and enhance functionality, have managing entities to oversee progress. Find an owner and hold him or her accountable. More on this later.

o   Build or acquire a framework

Even with lots of great experiences, insights and findings, without some kind of organizing structure, a collection of project post-mortems will pose an unwieldy and perhaps insurmountable barrier to leveraging collective lessons learned. Fortunately, PMI, ISO, ISACA, SEI and many other organizations have developed frameworks and a wealth of best practice information. I personally developed the Project Pre-Check Decision Framework to bring together the best of project management, management of change, software development and other practices and provide a Lessons Learned framework in my consulting practice. Select a structure you and your organization are familiar with and build it up with real world experiences and examples to facilitate effective use and enable real performance improvements. Or build your own framework.

o   Enforce usage

Having a comprehensive, easy to access, easy to use facility for mining best practices adds absolutely no value if no one uses it. So part of the Lessons Learned challenge, and a critical responsibility for the owner, is to ensure that everyone uses it, on every assignment, for every project. That use needs to involve a thorough review to identify and leverage applicable best practices and contributed learnings in a form and structure others can use. I can hear the groans already! More bullshit! More red tape! Get over it! If your organization wants to improve its ability to deliver major business and technology change successfully, a certain level of rigor and compliance is essential. What would you rather be, a PM with an incredible track record of successful project deliveries, enabled by an organizational ability to leverage lessons learned, or an incredible individual contributor with a mixed record of project successes? Your choice!

o   Manage the transformation of project experiences to organizational Lessons Learned

A key challenge is gathering the experiences and insights of each project participant and the collective wisdom of project teams and abstracting that information into the selected framework. Don't leave it up to the PM to post the project learnings to the Lessons Learned framework! The owner needs to assume that responsibility and establish the analytical mechanisms and quality and usability standards that will ensure that others, in different circumstances and on other assignments, can quickly and effectively gain value from the information.

o   Manage Lessons Learned value contribution

There is no point in doing anything beyond individual project post-mortems if you're not willing to manage the organizational value contribution. Getting a return from Lessons Learned means measuring usage, compliance, value derived and contribution frequency and identifying gaps and discrepancies in content and application. The measurement results provide the fodder to develop and implement continuing improvements to the practices, increasing the value delivered to the organization.

Where should Lessons Learned be managed? The PMO is an obvious suspect. The PMO mandate is usually very compatible with a Lessons Learned program. It typically has a broad view, encompassing enterprise or organizational initiatives, programs and projects. It should be tracking and reporting on aggregate project performance and taking steps to improve that performance, very compatible with a Lessons Learned program.

But, implementing a Lessons Learned practice across an organization is another change that has to be sold, prioritized, funded, initiated, staffed, managed and monitored. It is still very worthwhile but there is another option—the individual Lessons Learned practice. As a project management professional, you have undoubtedly committed yourself to a program of continuous improvement. An individual Lessons Learned practice starts and ends with you. You need to follow the same five steps reviewed above but you don't have to sell anybody else on the idea. You are the owner. You are the decision maker. As you gain value, and confidence, you'll build a track record and reputation others will notice. Also, share your experiences with your peers. Chances are you'll find an enthusiastic audience and perhaps an attentive sponsor in waiting. Good luck.

PART V
Decision Framework

- o **Change Domain**
- o **Environment Domain**
- o **Assets Domain**
- o **Project Domain**

The **FastPath** Decision Framework ensures that the factors that need to be considered for project success are addressed. It encourages a broad perspective when considering a change and its impact on an organization. It also provides the means to assess and rationalize a vast number of best practice opportunities into a manageable number of Decision Areas—the specific items that stakeholders need to consider and decide upon as a change progresses.

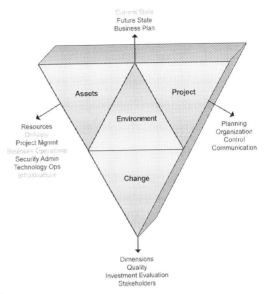

Figure 12—FastPath Decision Framework

Domains address the universal concerns that stakeholders must consider for each business and technology change: the Change itself, the Environment within which the change is being implemented, the Assets (resources, products and processes) that are available to support the change or that will be affected by the change, and the specific practices that will be used to deliver the change successfully. Factors are simply different facets of a Domain that help focus stakeholders' attention on more granular elements of the change. **FastPath** addresses all Domains but four of the eighteen Project Pre-Check factors are not included. Figure 12 shows the fourteen Factors that are included in bold.

If you think one or more of the excluded factors should be considered for a specific change initiative, go ahead and include them. Remember, both the full Project Pre-Check Decision Framework and the scaled down **FastPath** Framework are intended to be extended as appropriate to help stakeholders manage change successfully.

Decision Areas subdivide Factors still further to arrive at decisionable elements that represent and encapsulate proven best practices.

Stakeholders should visit each Decision Area as early as possible in the project life cycle, explore the relevance, the options and the impact. They must reach a consensus on a course of action for each Decision Area and ensure that the time and effort to enact each decision is reflected in the operational plan. Then they can proceed with confidence that critical ingredients which must be delivered, or are affected, or can be leveraged by the planned change have been identified and addressed.

Decision Framework

There are obvious dependencies among the Domains, Factors and Decision Areas. Project Pre-Check isn't concerned about these dependencies. Instead, the focus is on ensuring stakeholders make a call on each Decision Area at the appropriate time for the specific project. The Decision Areas provide the framework for an overall scorecard on stakeholder agreement. The sooner all Decision Areas are addressed and the decisions are agreed to, the better. The sequence of the decisions doesn't matter as long as stakeholders recognize and take into account the interrelationships in their decision making.

The diagram in Figure 13 shows the relationship of the four Domains to the eighteen Factors and the 125 Decision Areas in the full Project Pre-Check Decision Framework. The Decision Areas in bold are included in *FastPath*. You'll notice with some Factors that just the Factor name is in bold. In those situations (for example Quality and Investment Evaluation in the Change Domain), the whole Factor has been turned into a Decision Area. That reduces the number of stakeholder decisions initially required from a somewhat intimidating 125 to the more manageable 50 included in *FastPath*. Again, if there are specific Decision Areas from the full Project Pre-Check Decision Framework that you'd like to include in your *FastPath*, just add them to your questionnaire and reporting. What's important is that your stakeholders are focusing on the key decisions that matter to your project, whether it's 32, 84 or 147.

Each *FastPath* Decision Area is described in more detail in the pages that follow.

Project Pre-Check *FASTPATH* Decision Framework

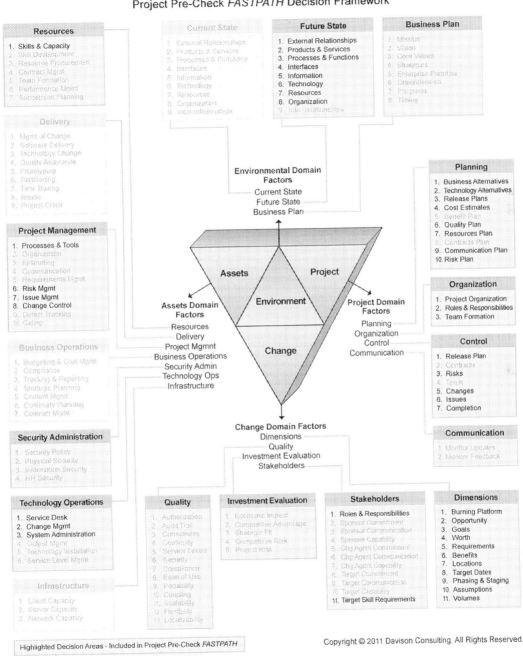

Figure 13—FastPath Domains, Factors & Decision Areas

Change Domain

"There is nothing so useless as doing efficiently that which should not be done at all."

Peter Drucker

The Change Domain encompasses the factors that enable the stakeholders to shape the planned change to deliver a responsive, cost-effective, quality solution that maximizes returns and can be completed within budget and on plan. Its focus is to ensure change parameters are fully defined, assessed and controlled in the context of sponsor, change agent and target expectations and capabilities.

The Change Domain addresses each project in relationship to four Factors: dimensions, quality, investment evaluation, and stakeholder capability.

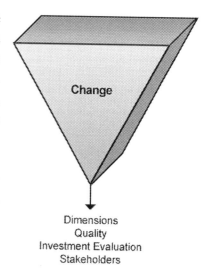

Change

Dimensions
Quality
Investment Evaluation
Stakeholders

Figure 14—Change Domain Factors

Dimensions Factor

The Dimensions Factor defines the breadth or opportunity to function. It should identify the business goals and objectives that will be addressed by the proposed change, the planned level of investment and expected benefits, the organizations and processes affected and the target implementation approach and time frames.

"Simplicity is the ultimate sophistication."

Leonardo da Vinci

The Dimensions of a change initiative are defined by the following Decision Areas: burning platform, business need or opportunity, business goals and objectives, worth, requirements, benefits, locations, target dates, phasing and staging, assumptions, success criteria and volumes, transaction mix and peak periods.

Stakeholders should be thoroughly familiar with each one of the Dimensions Decision Areas. They also need to understand the relationships among the Decision Areas and ensure a consistency of intent and direction that will provide a solid foundation for the deliberations and effort to come.

Burning Platform

The term "burning platform" is used by management of change practitioners to describe the current conditions or circumstances that would provide compelling motivation for an organization and its members to abandon the relative comfort of the current situation and venture into a new and perhaps risky future.

The term arose from a story about a fire on an offshore oil drilling platform where the workers chose to jump into the chilly ocean waves with flaming oil slicks all around rather than stay on the drilling platform. When asked why he had leaped into the ocean, one worker is reported to have responded "jumping into the ocean was possible death, but staying on the platform was certain death".

Stakeholders need to understand, articulate and communicate the "burning platform", or rationale, and the associated consequences that make the current state untenable. This helps to disarm the inevitable resistance and provides a shared frame of reference for constructive and focused action toward the target future state.

Business Need or Opportunity

Documentation of the Business Need or Opportunity should build on the burning platform and provide:

o   A description of the chain of events that led up to the planned change.
o   A statement of the problem being solved by the project.

o   Any other significant information about the project which will assist in understanding the need for the change.

o   A description of the relevant opportunities and risks.

Business Goals and Objectives

Clear and understandable goals and objectives are an essential pre-requisite for success. Robert Schaffer, in an article in the Harvard Business Review, stated "Without an ever-sharpening demand framework, improvement programs and activities are merely diversions from the real work of making our corporations more competitive"[21]

"You are what you measure." [22]

Hauser & Katz

According to an article by John R. Hauser, MIT Sloan School of Management and Gerald M. Katz, Applied Marketing Science, Inc., "every metric, whether it is used explicitly to influence behavior, to evaluate future strategies, or simply to take stock, will affect actions and decisions."[23]

Whether the metrics for a given change are tied into an enterprise based balanced scorecard or use basic change measurements like return on investment, budget, schedule, benefits, etc., metric selection is critical to the realization of the project's goals.

Selection of appropriate metrics helps crystallize and confirm stakeholders' expectations. It is a primary mechanism for communicating the expected results of a change for affected constituents and bringing about appropriate behavioral changes to achieve the desired results.

According to the former Meta Group, "Public performance measures enable high performers to mature and good performance measures enable high-performing teams to mature. Great performance measures enable the entire organization to rapidly mature".
Hauser and Katz suggest the following seven steps to establish meaningful metrics [24]:

1.   Start by listening to the customer
2.   Understand the job

3. Understand interrelationships
4. Understand the linkages
5. Test the correlations and test manager and employee reaction
6. Involve managers and employees
7. Seek new paradigms

The goals and objectives address the desired outcomes that will be achieved when the change is delivered successfully. Setting objectives should follow a structured process like the SMART format:

o   Specific
o   Measurable
o   Achievable
o   Relevant
o   Time-bound

Business Worth

Do you know what the project you're working on is really worth to the organization? If you don't, you're operating without one of the key sets of information needed to ensure project success. In the first of a new series on managing change initiatives, here is some guidance on how establishing worth—or affordability—can help you and your stakeholders improve project performance.

It is essential that the value of the change (what the organization can afford to spend) is stated explicitly in dollars and cents as early as possible. This will provide a framework for the identification and exploration of alternative solutions.

There are many ways to solve a problem, from multi-million dollar schemes that serve hundreds of thousands or millions of clients daily to remedies that cost a few thousand dollars and solve a problem for a few clients in the short term. Sometimes a fast, inexpensive implementation is the perfect solution when it isn't clear what the solution is and some real world experience is needed to gain valuable insights into the next steps. Sometimes the problems are thoroughly understood and a robust, fully configured but high cost solution is the answer.

Decision Framework

Let's look at an example of how you can facilitate the determination of worth for your project. An executive calls you into a meeting, or corners you in the hall, and proceeds to describe a major change initiative he or she has in mind. The dialogue goes like this: "You did a great job managing that ABC project for us last year and I'd like you to be involved in a new initiative we're planning."

The executive goes on to describe the venture and the wonderful benefits it will deliver to the organization and then concludes with "You did a great job of estimating and controlling the delivery of the ABC project within budget and on schedule. You're a pro! That's why I'm involving you! I need a ballpark number for our new project. Don't worry, we won't hold you to it but we just need a rough cut of the costs and timeframe for our pitch to the Executive Committee on Friday."

Sound familiar? What should you do? Give the executive a ballpark time and cost? Ask for more time to understand the change and come up with a reasonable estimate? Tell the executive that you'll need *n* weeks or months to assemble a team and determine the requirements before you can develop a reasonable plan and costs? All of the above?

Here's an alternative approach—when the executive asks for the ballpark numbers, take this approach: "Well, how much can you afford?"

Executive: "What do you mean 'how much can I afford'? I'm just looking for a ballpark number here!"

You: "Well, the amount you can afford to spend will influence the alternatives we consider and that will be a major contributor to time and cost. For example, if the change is worth $100K to the organization, we won't even consider million dollar solutions."

Executive: "So, how do I go about figuring out what the change is worth? We have the Executive Committee presentation on Friday!"

You: "There are three things you need to consider to arrive at a value for worth:

1) what's the projected bottom line impact that you're hoping to achieve,
2) how quickly do you need to reach the payback point, and
3) are there other changes, planned or in progress, that you're dependent upon or competing with.

Once you've determined the worth, you can use that number as a proxy for the cost and indicate to the Executive Committee that you'll be evaluating alternatives that can be delivered within that number."

Executive: "Sounds reasonable! Can you give me some time to help come up with a figure for worth I can live with?"

You: "Of course!"

What just happened here? A number of very positive forces have been set in motion:

o   The executive is about to commit to a target figure and timeframe for the project. That worth figure belongs to the executive, usually the initiating sponsor. It will be used for the duration of the project to influence and direct every stakeholder decision. The project manager will still be on the hook for estimating and delivering their part of the change, but within the context of worth.

o   Worth provides a framework for the identification and exploration of alternative solutions. There are many ways to solve a problem, from multimillion-dollar schemes that serve hundreds of thousands of clients daily to remedies that cost a few thousand dollars and solve a problem for a few clients in the short term. Without this statement of worth, or affordability, considerable time, effort and energy can be wasted exploring options that will not be viable or appropriate to the circumstances.

o   Worth can also be expressed in more granular terms to give greater guidance to solution identification and assessment:

   •   Investment worth is the amount the organization is willing to spend to achieve the anticipated benefits

   •   Operating worth is the maximum increase or decrease in annual operating costs (business, technology, resource costs, etc.) that the organization is willing to incur to deliver the anticipated benefits.

o   Worth provides a vehicle for managing scope change requests. No longer can the requesting stakeholders abuse the project manager for escalating time and costs. With worth, the change requestor now has to convince the other stakeholders that the requested change delivers sufficient value to justify any change in costs or schedule.

Decision Framework

Tom DeMarco, a recognized leader in cost forecasting and president of Price Systems, spoke about the findings reported in a Price study (see "A Cracked Foundation"). The study found, among other things, that government IT executives believe that almost half of all failed projects could be avoided if project baselines were more realistic. What better way to deliver more realistic project baselines than to frame them with a fully supported statement of worth?

Remember to ask the question, "What is this change worth?", and make sure you get answer that all stakeholders acknowledge.

Requirements

IBM's Unified Process (UP) defines a requirement as "a condition or capability" to which a solution must conform. The Standish Chaos studies have consistently ranked the need for clear, comprehensive requirements as one of the top ten factors contributing to project success. Most other studies and commentaries also attribute project success, in whole or in part, to the requirements area.

Stakeholders need to pay considerable attention to and agree with the specific requirements for a change. The change could be business or technology driven, an organizational change initiative, merger and acquisition activity, a new product or service, a process improvement effort or any of the thousands of other changes modern organizations need to deal with. It doesn't matter. Stakeholders need to be on top and on side.

Paul Glen, an author and principal in C2 Consulting, summarized the challenge and the opportunity in a recent article for Tech Republic:

> *"We should think of a set of requirements as being like a multilateral treaty among a group of nations. Representatives of nations negotiate treaties by seeking out points of agreement, acknowledging constraints, compromising and trading off. The forging of a treaty is an active and difficult process of creation. No one would suggest that you 'gather paragraphs' to write a treaty.*
>
> *So we must negotiate requirements among the many stakeholders whose positions and interests need to be acknowledged. There are the business interests of executives who fund projects, of course. There are the utility needs of the users who*

will work with the systems every day. There are also the technical needs of those who build, deploy and support those systems. The list can go on and on.

Successful projects begin not with a harvest, but with a difficult set of discussions on what should be done. If you stop trying to gather requirements and start negotiating them, your projects will yield richer crops."25

One useful acronym for defining requirements is another use of the SMART acronym, described by Nick Jenkins in A Project Management Primer[26]:

Specific	A goal or requirement must be specific. It should be worded in definite terms that do not offer any ambiguity in interpretation. It should also be concise and avoid extraneous information
Measurable	A requirement must have a measurable outcome. Otherwise you will not be able to determine when you have delivered it.
Achievable	A requirement or task should be achievable. There is no point in setting requirements that cannot realistically be achieved.
Relevant	Requirements specifications often contain more information than is strictly necessary which complicates documentation. Be concise and coherent.
Testable	In order to be of value requirements must be testable. You must be able to prove that the requirement has been satisfied. Requirements which are not testable can leave the project in limbo with no proof of delivery.

Table 8—SMART Practice for Requirements Definition

Benefits

Well-articulated business benefits (quantitative and qualitative) are a pre-requisite for benefit delivery and can include:

- o Revenues
- o Operating costs
- o Market share
- o Future budget and cost center impact
- o Customer satisfaction
- o Customer retention
- o Customer penetration

    o   Product or service quality

However, project benefit statements and forecasts often don't cover the important factors that can help get target audiences on side. That can include improvements in the working environment, greater opportunities for staff, chances for personal financial gain, etc. One approach for articulating the expected benefits to reflect the different frames of reference involved in a change is through the use of stories.

A story is a most powerful tool when it comes to articulating one's views, hopes and desires and helping others understand and embrace the opportunity. Most of us can recall the power of an anecdote, told by a teacher, manager or leader that enhanced our interest in and understanding of the subject matter being presented.

Stories can help others understand the nuances of a major business or technology change in terms that relate meaningfully to each individual. Stories can be used to share valuable wisdom, lessons learned, and how-to's that can contribute to personal insight, comprehension and action supporting the planned change.

As important, however, the stories should reflect the view from each participant's frame of reference: sponsors, change agents, each unique target audience and champions. It's an opportunity for each stakeholder to give their own personal views and insights, to reflect on and recall relevant experiences, to include anecdotes that increase understanding and buy-in.

Locations

The number of physical locations involved in developing a solution and affected by solution delivery can have a significant affect on costs and pose incremental risk to a project.

Typically a major business or technology change will affect a variety of different locations: in different countries, states, provinces and municipalities, at various client, partner and supplier sites, at head office, branch offices, home offices and homes.

Therefore, it's important to understand and address the location challenges early in the change cycle and develop approaches to minimize the costs and risks. Identifying the variety of locations and the common and unique characteristics of each is necessary for managing the impact of the change successfully over time.

Targets Dates

It is important to understand and clearly articulate the business rationale behind key target dates to ensure appropriate decisions are taken as the project progresses. Target dates can represent responses to a variety of internal and external pressures:

- competitive
- financial/budgetary
- legislative
- risk management
- strategic positioning
- executive whim

Imposing target dates that have no relationship to real external or internal needs can significantly increase costs and introduce unnecessary risks. Similarly, failure to articulate and communicate the reason behind critical targets can expose an organization to significant loss. When all stakeholders understand the critical drivers behind a target date, it gives everyone an opportunity to identify and vet creative options to help achieve the project's goals while constraining costs and risks.

Phasing & Staging

Abrams's Advice: When eating an elephant, take one bite at a time.

Your organization is undertaking a major new project that will have a significant impact across the board. If we know anything about project risk, we know the bigger they are the harder they fail. What can you do to improve your chances of success? Here are some approaches to consider that will ensure your organization gets value out of the undertaking.

The September, 2006 issue of Baseline Magazine included an article on the *Top 10 Project Pitfalls You Can Avoid*. Number 1 on the list? *A project's scope is too monolithic and gargantuan*. Here's the example they used:

> *In 2001, McDonald's planned to spend $1 billion over five years to tie all of its operations into a real-time digital network. Eventually, executives in company headquarters would have been able to see how soda dispensers and frying machines in every store were performing, at any moment. But after just two years, the fast-food giant threw in the towel.*

Decision Framework

There are, unfortunately, thousands of other recent examples of projects that were too big to succeed and lots of analysis to show us why jumbo projects don't make sense.

A study by the Standish Group revealed that projects costing less than $750,000 succeed 55 percent of the time, those in the $1 million to $2 million range have an 18 percent success rate, and those in the $5 million to $10 million range succeed only 7 percent of the time.

Ford Motor Co. lost an estimated $220 million on installing an electronic purchasing system. When Ford launched a procurement system dubbed Everest in 2000, the game plan was simple: exchange information on orders, accounts receivable and inventory status with suppliers electronically. The system was announced in 1999 and dismantled in August 2004. Ford pulled the plug when it realized more heavy investment was needed. The big issue: Ford's suppliers were already tied to the automaker's systems via electronic data interchange applications and didn't see a benefit in changing to Everest.

The Internal Revenue Service's failed $4 billion Tax Systems Modernization Program (TSM) of the 1990s is a glaring example of this tendency to bite off too much. TSM failed because it took a "big bang" approach in which the IRS tried to build a new and gigantic information system to integrate many disparate and out-of-date systems.

Having learned from this lesson, Arthur Gross, former CIO of the IRS, declared that the agency would from now on take "an evolutionary approach to modernization," building on systems already in place.

Identifying phasing opportunities (how a solution will be built or assembled) from a business perspective is essential for creating change plans that optimize business value and manage risk. Because the risk of failure increases with size, stakeholders should considering phasing mandatory.

Phasing can be done by:

- customer or customer segments
- product lines or features
- geographically
- process or function
- risks or exposures

One technique for establishing priorities and phasing options is to use the MuSCoW technique where each element of the change is assessed according to the following categorization:

Must—Minimum useful subset of business functional and structural requirements, features, locations and markets. Must be done in order to do business

Should—Workaround, but costly. Include as many of these in the first release as possible. Required by the business but there is a workaround that can be used temporarily.

Could—Great benefit but can manage without it for some time. Added benefit to the business but can be deferred.

Would—Wait until higher priority requirements are complete. Added benefit but needs to wait until higher priority requirements are done.

Staging relates to the options for managing the rollout or implementation of a solution. Avoiding the "big bang" implementation and staging smaller, targeted implementations can help improve response times, manage risk and enhance value delivery. The options for staging are similar to the phasing options outlined above.

Phasing and staging decisions are often difficult calls to make—too many and the costs, complexity and momentum can be negatively influenced, too few and responsiveness and return can be sacrificed and risks magnified. It's also important when establishing priorities to consider all of the elements that go together to make up an acceptable implementation package. Sometimes a few "coulds" or "woulds" can help add excitement to boost enthusiasm for and commitment to the change.

Given the risks of a project too big, ask the stakeholders how they would like to see the change phased and staged. Given the Standish findings mentioned above, "we need to do it all at once" is NOT an acceptable answer. And when pushed, you'd be surprised at the kinds of creative options stakeholders can come up with.

Assumptions

Tylk's Law: Assumption is the mother of all foul-ups.

Decision Framework

Changes are initiated and carried out in response to a number of drivers. Often these drivers are based on assumptions made by the initiating sponsors or reflect perceived constraints or pressures. The driving assumptions and constraints need to be fully articulated, acknowledged and validated by the stakeholders and monitored throughout the course of the project. Any opportunity to minimize dependence on assumptions should be embraced.

Changes in the status, breadth or depth of the assumptions and constraints should be monitored and evaluated to understand the impact on decisions already taken in the four Domains. Decisions should then be adjusted accordingly.

Volumes, Mix and Peaks

Many changes in the business and technology world are undertaken explicitly to drive or support increased volumes; more clients, introduction of new services, greater use of existing services, anything that leads to more business. Even small changes, such as an advertising campaign, can drive huge changes in usage. Here's some guidance on what you should be looking at with respect to anticipated volumes and how that insight can help you and your stakeholders improve project performance.

Most projects will affect a number of business or operational activities whose changing volumes will, in many cases, be a reflection of the success of the change. The activities can range from business transactions to number of clients, web site hits, number of projects, Help Desk calls, staffing hires and departures, number of sales, number of contract terminations, number of widgets produced, etc.

The projected volumes, mix and anticipated peak periods associated with these activities and the levels of confidence behind these projections can have a dramatic impact on the nature of the solutions considered and the associated costs and risks. Therefore, it's important to identify the activities most reflective of the change's impact and desired end result, establish high level estimates of volumes, mix and peak period profiles and agree on the level of confidence for each forecast. This will help ensure that subsequent analytical and design activities are focused and relevant.

Consider **The London Stock Exchange Big Bang**—In 1986, the London Stock Exchange introduced a number of fundamental changes to their operating rules and processes (known as the Big Bang) including the introduction of automated trading. Just before the changes were implemented, the average number of daily trades at the London Stock Exchange was 20,000, amounting to about £700m worth of shares changing hands.

After the introduction of automated trading the figure went up to a daily average of 59,000 trades a few months later, almost triple the pre-change rate. During October 2008, over seventy-five billion shares were traded on the London Stock Exchange. That's almost 4 billion trades a day, quite a leap over twenty years.

Web sites give us examples almost daily of the "Oops!" factor, where organizations don't do a very good job of predicting appropriate volumes for new programs or services and end up with disappointed and disgruntled users, many of whom go elsewhere to get their products, services and information. Some embarrassing examples:

Red Cross Tsunami Website Crashes—in December, 2004 a Red Cross website to help anxious relatives locate survivors of the Indian Ocean tsunami disaster partially crashed after being overwhelmed by some 650,000 hits in its first 24 hours.

Wal-Mart Website Crashes on Black Friday—On November 24, 2006, high traffic disrupted Wal-Mart Stores Inc.'s web site for much of the day. Black Friday, the day after the Thanksgiving holiday, marks one of the year's busiest days for retailers and the official start of the holiday shopping season.

The troubles came a day after Amazon.com Inc.'s site had brief disruptions because of a Thanksgiving Day sale on Microsoft Corp.'s Xbox 360 video game machines.

For much of Friday morning, attempts to open Walmart.com resulted in blank pages, delays or other problems. By early afternoon, visitors were simply told to come back later. A Walmart.com spokeswoman blamed a "higher than anticipated traffic surge."

O2 Website Crashes Under 3G iPhone Demand—in July, 2008, O2 customers determined to get their hands on apple's new 3G iPhone caused the network's website to crash as they rushed to pre-order the new device. Pre-orders opened at 8am and within an hour the website had buckled under what an O2 spokesman described as phenomenal demand for the new iPhone.

UK Government's National Pandemic Flu Website Crashed—On July 23, 2009 the new national pandemic website crashed within four minutes of going live. The site gives people information about swine flu, and has a series of questions to enable them to diagnose themselves. The problem came in spite of prior assertions by the government that the site would be thoroughly tested, and that any glitches would be viewed as "outrageous".

There are thousands of examples of volume related disruptions. Chaos from volume swings can affect anyone; private companies, government agencies, not-for profits, big, medium and small. Also, inordinately low volumes can be as devastating as unanticipated high volumes. I recall one organization that spent over $6 million upgrading processes and systems to support a new financial product only to end up selling a grand total of 20 over 2 years. There were some very viable, far less expensive options available to support that level of activity.

So, remember to ask the question "what will this change do to business volumes, the mix of transactions that will result, the peak periods we'll experience and what is our level of confidence in these forecasts". Make sure you get an answer all stakeholders agree with. That information is essential for identifying and implementing alternatives that are appropriate to an organization's needs.

Quality Factor

When we talk about producing a "quality" product, what do we actually mean? Is it easy to use? Fast enough? Flexible? Does it work the way we want and produce the results we expect? Consistently? Here are some features to consider when trying to address the meaning of quality for your project.

In many cases, lack of project success can be attributed, not to budget and schedule failures, but to oversights and omissions in the features, functions and capabilities delivered. The quality factor identifies the expected levels of performance for the planned change across the entire operating spectrum. The quality needs and expectations are used to shape the planned solution from the initial phases through to and after implementation.

The quality features mentioned here have been adapted from the Quality Assurance Institute and other sources and provide a framework for analyzing and establishing stakeholder expectations and priorities on a number of critical fronts.

Quality Features

For each quality feature, it is important to establish early in the change life cycle specific, measurable goals to which all stakeholders subscribe. In fact, multiple goals for a given factor may be required to reflect differing stakeholder needs. In addition, stakeholder agreement on Decision Area priorities must be established to understand potential conflicts and direct follow-on efforts.

Authorization

Authorization means assurance that data is processed in accordance with management's intent. There is both general and specific authorization for the processing of business transactions and data access.

For example, NASA's 80,000 workers are actually a minority of the computer users with an account on some space agency system. There are also the scientists who log in remotely from universities around the world, as well as corporations and foreign space agencies that partner with NASA on specific projects. Because the creation of these accounts wasn't tracked systematically, NASA was able to offer only a "best guess" at the total number of users: about 275,000.

Audit Trail

Audit trail refers to the capability to substantiate the processing that has occurred. The processing of data can be supported through the retention of sufficient evidential matter to substantiate the accuracy, completeness, timeliness, and authorization of data.

For example, Diebold Election Systems admitted in a recent California state hearing that the audit logs produced by its tabulation software miss significant events, including the act of someone deleting votes on election day 2008.

Correctness

Correctness refers to the degree to which the data entered, processed, and output is accurate and complete. Accuracy and completeness are achieved through controls over business transactions and information. The control should commence when a transaction is originated and concludes when the transaction information has been used for its intended purpose.

Decision Framework

You may have heard the story about a programmer at a financial institution who changed the interest rate calculation to three decimal places, deposited the amount from the third decimal from every interest calculation into his account and the first and second decimal amounts into the clients' accounts. Everything balanced! But was it correct?

Continuity of Processing

Continuity refers to the ability to sustain processing in the event problems occur. Continuity of processing assures that the necessary procedures and backup information are available to recoup and recover operations should the integrity of operations be lost due to problems. Continuity of processing includes the timeliness of recovery operations and the ability to maintain processing periods when the business process is inoperable.

Service Levels

Service levels mean assurance that the desired results will be available within a time frame acceptable to the user, including response times, availability (e.g. 24x7x365) and consistency. To achieve the desired service level, it is necessary to match user requirements with available resources. Resources include technical capabilities (input/output, communication facilities, processing, and systems software capabilities) as well as people, facilities and a variety of logistical and enabling services.

Security

Security means assurance that resources will be protected against accidental and intentional modification, destruction, misuse, and disclosure. The security procedure is the totality of the steps taken to ensure the integrity of assets from unintentional and unauthorized acts.

Compliance

Compliance means assurance that a solution is designed in accordance with organizational, regulatory and legislative methodologies, policies, procedures, and standards. These requirements need to be identified, implemented, and maintained in conjunction with other requirements.

Ease of Use

Ease of use refers to the amount of effort required to learn, operate, prepare input for, and interpret output from a business process. It deals with the usability of the process to the people interfacing with the process. Each interface, process and user can have a different ease of use need.

For example, after Bank of America required online customers to use a new log-in mechanism to thwart phishing, calls to the bank's service centers climbed by 25%.

Portability

Portability refers to the need for and effort required to transfer a process or data from one environment to another, for example, from a head office environment to client sites, or a disconnected notebook computer or handheld device. The effort includes data conversion, program changes, operating system, and documentation changes.

Coupling

Coupling refers to the need for and effort required to interface one process or technology with all the other processes or technologies which either it receives data from or transmits data to.

Scalability

Scalability refers to the ability of a process or system to grow beyond the initial target volumes and continue to meet all other quality targets. Scalability includes both manual and automated components, covers both peak and average demand and should address all pertinent factors including number of users, transaction volumes and mix, data volumes and physical locations.

Decision Framework

Flexibility

Flexibility refers to the extent to which future changes in business function or technical capability can be made without time-consuming, complex or costly changes. Flexibility includes the effort, time, cost and processes necessary to deliver the expected change and satisfy the other Quality Factor targets.

For example, Ontario's welfare system was re-designed by a major consulting firm for hundreds of millions of dollars and apparently the upgrade didn't even allow for a rate increase. Oops!

Localizability

Localizability (or internationalization) refers to the need for many solutions to adapt to a variety of languages, cultures and conventions. Many of the above factors will need to be assessed in this context and any unique requirements addressed.

That's a lot of stuff to cover but it's information that can be critical to ensuring a successful change. So, remember to ask each stakeholder about their needs and expectations on each of the quality factors. Initially you may just get blank looks. But as each stakeholder ponders the factors, you'll build a collective and comprehensive understanding of how the factors relate to the planned change. Just make sure that all stakeholders agree with the collective view you get.

"Quality is not an act. It is a habit."

Aristotle

Focusing on quality, building it into the solution as it's being conceived and delivered, and testing from inception to completion can be the difference between success and failure. Consider the following example where a passion for quality, along with tenacious leadership and the support of an effective stakeholder group, turned a failing organization around.

Tenacity Can Achieve Miracles

The Situation

This provider of billing and customer information services to the utilities industry was experiencing significant customer dissatisfaction with the quality of its applications and services, contributing to loss of customers and a serious revenue decline.

Software updates and releases were handled by the internal development group. Responsibility for quality was shared between the programmers on each project and a central quality control unit. There were no standard development or quality practices. The development teams relied upon previous approaches used on a specific software package or service. The sales organization, which drove much of the enhancement activity, placed much greater emphasis on time to market, which was a significant contributor to the quality problems.

The CEO challenged the CIO to turn the situation around quickly through an order of magnitude improvement in the quality of delivered software and services.

The Goal

The CIO recognized that a number of changes would need to be made to their practices to achieve the CEO's mandate.

He proceeded to hire a manager to head the quality assurance function and guide the other changes. She had a wide range of experiences and accountabilities and an impressive track record. Her mandate was to deliver the changes within an 18 month time frame and achieve the targeted order of magnitude quality improvements to increase customer satisfaction and reverse the revenue decline.

The new QA manager took over a team of 32 staff, half of them located centrally in head office, the rest in four other offices spread across the country plus a small offshore testing facility in India. She was also responsible for supporting 60 applications and services for over 50 clients, most being large, very demanding utilities.

Her first week on the job was spent talking to the senior managers and staff to get their thoughts on what problems existed and how best to tackle them. She discovered the following views:

- There was inadequate documentation on the core systems and services, inadequate documentation about business requirements and application functionality for new offerings and a mishmash of project management practices to guide the projects.

Decision Framework

- There were frequent and ad hoc changes to the planned new releases.
- There was an absence of standard project management, development and quality practices which hampered the movement of staff to priority projects and created conflict and ill will when dealing with other organizations and clients.
- There were no controlled test environments for the core applications and services and no management and reuse of test conditions, cases and scripts.
- There were ongoing communication gaps among the development groups, Quality Control, Computer Operations, Sales and the clients.
- There was no or limited technology available to support requirements traceability, test planning and execution, release builds and promotion to production

She did a stakeholder map to identify who her key partners in this endeavor would be. They included:

- The CIO, who identified himself as sponsor of the change.
- The VP Sales, a key target because of the relationship with the sales staff and their clients.
- The VP System Development, also a key target because of the planned changes to development and quality practices
- The VP Computer Operations, an important target because of the planned changes to improve the build and promotion processes and the need for more responsive client support.
- The clients, for obvious reasons.
- Herself, manager of QA, as the change agent and also an important target because of the changes she would have to implement in her own organization

With the exception of the client representation, this became her initial stakeholder group. The Sales VP would wear two hats, as a proxy for the clients and representing the Sales organization. All material decision would be reviewed and approved by these players. All decisions needed unanimous consent.

Her next step was to develop and vet the vision for the changes that were needed, including the sequence of implementation and the planned time frame. When she had obtained senior management buy in, she launched her project. Her budget was just over $800,000 including software.

The Project

The first order of business was to communicate the vision to all those who would be affected by the upcoming changes. The Sales VP and Operations VP cooperated fully and encouraged their managers and staff to listen, provide feedback and get on board. Her vision called for a first wave of process improvements including the project management, development and QA processes, a testing technology project and a metrics and reporting initiative. These were to be followed by a second wave including release coordination, configuration management and document management

The Development VP refused to return her calls and didn't attend her meetings. She went below him to a Development manager who had taken considerable heat for quality issues on his applications and had expressed a desire to be an early adopter of the new practices.

She formed a team with representatives from the Development manager's group, Operations, Sales and QA to evolve the project management, QA and development processes. The mandate to that team was to use the best in house methods available and beef them up with industry best practices, from PMI, SEI and QAI among others, in a six week time box. That work was completed in the targeted six weeks and included high level documentation, references to external best practices, examples where possible and general training materials.

She appointed an experienced QA lead to manage the implementation, starting with the supportive Development manager's team and two projects just getting underway. As those two projects progressed in applying the new methodologies, gaps were identified and plugged, errors and omissions were fixed, new examples were collected and training materials and methods were updated.

On the testing technology side, the QA manager had been through a formal technology assessment, selection and implementation process in her last job. She was very familiar with the available offerings. In the interests of time, based on that experience, she pulled together a formal recommendation including assessment of the available alternatives and specific recommendations and reviewed it with the stakeholder group. All except the Development VP approved the proposal. In spite of the requirement to have unanimous agreement on all stakeholder group decisions, the CIO (the sponsor of the initiatives) gave approval for the QA manager to proceed with the technology acquisition and implementation and indicated he would work with the Development VP to resolve his concerns.

Decision Framework

The QA manager proceeded to form another team, led by another of her QA leads, to implement the technology and develop standards and practices and then apply those to supporting the two development projects. The team included staff from Operations, QA and Development (from the supportive Development manager's group). She was also able to get a staff member from the client involved with the two development projects.

On the metrics front, she consulted with the CIO and VP Sales about their views on the metrics required and, based on those discussions, recommended three initial measures: customer satisfaction and the change from period to period; quality, as measured by the number of critical system defects (failures in production) and IT productivity, as measured by revenue per staff. The last metric, site productivity, was an attempt to measure revenue growth in conjunction with improvements in IT productivity. With the exception of the Development VP, the stakeholder group approved the recommendations. The QA manager was given the green light to proceed.

After two months on the job, the QA manager had built and sold a vision and developed, pitched, received approval for and launched three key initiatives. She communicated formally on a weekly basis up, out, down and sideways. She also engaged anyone who wanted to talk about the program in the form and timeframe appropriate to the need. And, she took it upon herself to monitor the grape vine, to see what people were thinking, feeling and saying.

While all this was going on she also executed her duties as manager of the QA unit and helped her staff not formally involved in the three projects get up to date and on side.

The Results

The three initiatives were extremely successful. On the process initiative, the first two projects were delivered with zero critical defects and slightly beyond the initial target dates. However, the client was thrilled, to some extent due to the involvement of their staff in the testing technology undertaking. The testing technology project worked with the two project teams and the process group to build and reuse test scripts to cover the changes being made, resulting in much better test coverage, less time to execute, improved productivity and better quality. On the metrics front, the tie-in of the three metrics to IT staff bonus calculations and publishing the monthly results throughout the company had a palpable effect on the company culture. The CEO noted the effect in one of his quarterly updates.

The first wave of changes was rolled out across the organization in fourteen months. The second wave was completed in an additional ten months. Here's how the actual results looked:

Metrics	Before Changes	After One Year	After Two Years
Customer Satisfaction (1-Very Dissatisfied 10-Very satisfied)	4.1	6.3	8.4
Critical System Defects	41	3	0
IT Productivity (Revenue/FTE)	$48K	$55K	$81K

How a Great Change Agent Helped

This is an interesting study. Obviously, all the initiatives clearly qualify as projects. Yet the QA manager didn't really run them according to the usual model, with clear requirements, sign-offs, prescribed phases, etc. There was no risk plan, no issue log, no formal change control, no test plan, no detail project plans.

What allowed the QA manager to achieve these stellar results? I think five factors contributed to her success:

Decision Framework

- The QA manager had the knowledge, skills and capabilities to lead the initiative. She understood enough about project management and software development to understand the relationship to quality and productivity. She had a deep understanding of quality assurance, quality control and supporting technologies. She had the knowledge on tap to create the vision and the plan and get it approved.
- She knew, either instinctively or formally, how to manage change. She pulled the stakeholder group together. Instead of waiting for the Development VP to get on side or get lost, she went below him to a supportive and needy Development manager. She knew she could get away with the move because she had the support of the other stakeholders. She leveraged the burning platform—declining revenue and customer satisfaction—to get the decisions and resources she needed.
- She was a confident and gifted communicator. She communicated frequently, to all affected targets. She listened and acted on the messages she received. She involved the client, including bringing one client into the stakeholder group in the second year.
- She knew how to form effective teams—small groups of staff with clear mandates, appropriate availability, the right perspectives, knowledge and skills and short term time frames to deliver an actual result.
- She acted! She didn't wait around for someone else to tell her what to do. She didn't wallow in analysis paralysis. She went out and got stakeholder approval. She took clear, decisive steps. She did what she said she was going to do. When things went wrong, and they did, she was the first to acknowledge a problem and collaborate with those affected to seek an acceptable resolution.

One final note: the Development VP, who opposed the QA manager's plans and refused to participate, was replaced.

Investment Evaluation Factor

Do you really know why the project you're working on has received scarce organizational capital? Are the reasons for undertaking this particular initiative versus another clear to all? Here is a framework for making sure everyone understands where the project fits in the scheme of things.

Changes are conceived and initiated for a multitude of compelling reasons. However, most are launched in response to four key factors: economic impact, competitive advantage, competitive risk or strategic fit. Investment evaluation assesses the rationale for change, the level of risk and a project's priority relative to other possible undertakings. The investment evaluation approach can be used to assess projects competing for an enterprise's capital and resources in an annual or quarterly planning cycle and can be used to shape a portfolio of projects that most effectively addresses the organization's needs. A similar concept is included in the Val IT Framework, developed by the IT Governance Institute to help organizations optimize value from IT investments. It includes as a key management practice "Evaluate and assign a relative score to the programme business case".

Suggested levels and scores are provided below to provide a means of comparing change initiatives on a quantitative basis. Organizations can also assign weights to the factors and change the scores allocated to reflect necessary changes in emphasis from period to period.

Organizational Value

Organizational value encompasses four views on the motivations for and returns provided by a given change initiative. The source for the material presented here was forgotten long ago but our indebtedness to the author remains. The practice has been refined over the years and has proven effective for one project, for a couple of competing projects or for an annual portfolio planning program. You choose. The four views; economic impact, competitive advantage, strategic fit and competitive risk, are described below.

Economic Impact

Economic impact considers the value expected to be delivered to the organization in hard dollars (tangible) and soft dollars (intangible) as expressed in terms of payback, return on investment, internal rate of return or some other similar standard.

The expected benefits and costs will evolve over the course of the project as the scope, priorities, issues and risks are managed to meet organization needs. So the assessment should be revisited over the course of the change as conditions dictate. The initial cut at the figures can use initial projected benefits and the statement of worth or affordability as a proxy for costs until the scope has been solidified and reliable estimates have been developed. The suggested rankings for economic impact include:

Payback Period—Hard Dollars	Value
under 1 year	5
1-2 years	4
2-3 years	3
3-4 years	2
4-5 years	1
5+ years	0
Payback Period—Soft Dollars	**Value**
under 1 year	3
1-2 years	2
2-3 years	1
3+ years	0

Table 9—Economic Impact Values

Competitive Advantage

Competitive advantage focuses on the value derived from a new business strategy, product or service. The factor provides an avenue for expressing the window of opportunity for a new initiative in the overall assessment of the change. The suggested rankings for competitive advantage include:

Competitive Advantage	Value
Greatly improves competitive position by providing a level of service/products unmatched by competitors (24 month window)	5
Substantially improves competitive position by providing a level of service/products significantly beyond most competitors (12 month window)	3
Moderately improves the competitive position by providing a level of service/products beyond most competitors (6 month window)	1
Brings the level of service/products in line with the competition	0

Table 10—Competitive Advantage Values

Strategic Fit

Strategic fit focuses on the degree to which a project supports or aligns with stated corporate and line of business strategic goals. The strategic fit factor provides an avenue for enhancing the priority of innovative or alignment applications that are in direct support of business goals and strategies.

Strategic Fit	Value
Critical to the success of the line of business and corporate goals	5
Necessary to deliver the line of business and corporate goals	4
Supports the line of business and corporate goals	3
Tactical with strategic component	2
Tactical solutions only	1
Has no direct or indirect relationship to the corporate vision	0

Table 11—Strategic Fit Values

Competitive Risk

Competitive risk assesses the degree to which failure to do the project will cause competitive damage. The competitive risk factor provides a means of recognizing the relative importance of an initiative that is being launched in response to a competitive situation and that cannot be fully justified on a payback basis or fully support business goals and strategies. The suggested rankings for competitive risk, on a score of 0 to 5, include:

Competitive Risk	Value
Postponement of the project will result in irrevocable damage	5
Postponement of the project will result in further competitive disadvantage	4
Postponement of the project may result in further competitive disadvantage	3
Can be postponed for up to 12 months without affecting competitive position.	1
Can be postponed indefinitely without affecting competitive position.	0

Table 12—Competitive Risk Values

Decision Framework

<u>Project or Organizational Risk</u>

Project or organizational risk focuses on the degree to which the organization is capable of carrying out the changes required by the project.

Project or Organizational Risk	Value
1. Magnitude of change	
Major change for targets	5
Moderate change for targets	3
Minor change for targets	1
2. Primary target of change	
Customer	10
Field or remote locations	5
Internal	1
3. Management commitment	
No active sponsorship	10
Actively managed at mid-management levels	3
Actively sponsored at highest level	1
4. Project management/skill sets	
Inexperienced	10
Somewhat experienced	3
Experienced and capable	1
5. Cross-organizational implications	
Affects all services or lines of business	10
Affects multiple services or lines of business	3
Confined to limited services or lines of business	1
6. Magnitude of business & technology changes	
Major change to existing environment	5
Completely new business or technology	3
Minor changes to existing environment	1
7. Clarity of business need	
Ill defined	10
Generally understood	3
Clearly defined	1
8. Knowledge of target solution	
Brand new product or service	5
Somewhat new/limited applications	3
Very well known industry wide	1

Table 13—Project Risk Values

o   For total scores of 8-26, the risk rating is Low

o   For total scores of 27-46, the risk rating is Medium

o   For total scores of 47-65, the risk rating is High

The suggested categories and corresponding values for project or organizational risk are reflected above. The overall risk rating for a change can be established by totaling the values and comparing the results to the scale.

This risk matrix can be used to assess each project independently or to compare the risk profiles of a number of projects in the development of a portfolio to ensure a suitable overall risk load. It can also be used to assess the risks associated with a number of possible courses of action for a given change.

Stakeholders Factor

Organizations make many demands on executives and managers regardless of whether they are in a business or technology leadership role or the public or private sector. However, the success of a change initiative is almost always dependent on the vision and passion of stakeholders and their ability to commit the time and effort required to see a change through to full completion.

Therefore, it's essential that stakeholders are identified, understand their roles and responsibilities and are fully engaged. Without that engagement their ability to commit to a planned change in light of other demands may conflict with the change or divert attention and jeopardize project success.

The level of commitment, the effectiveness of communications and the capability of stakeholders are vitally important indicators of implementation success. High levels of commitment are especially critical from the senior executive who has the organizational power to authorize and legitimize the change. Similar commitment must be exhibited by every other manger in the organizational hierarchy down to the ultimate target for the change. Such a pattern is required to reinforce the importance of the change at each organizational level. Commitment must be demonstrated by effective communication and by actions that reinforce what is communicated.

Decision Framework

"The quality of a leader is reflected in the standards they set for themselves."

Ray Kroc—founder of McDonald's Corporation

Stakeholders need to assess objectively sponsor, change agent, target and champion commitment, communication and capability on an ongoing basis. It means looking around the table at the members of the guiding coalition, including oneself, and ensuring that all are comfortable with and confident in the collective abilities to get the job done.

The Decision Areas for Champion commitment, communication and capability have not been included in the Decision Area catalogue because these are the qualities that would be considered in appointing a Champion in the first place. The other roles, on the other hand, are most often a function of one's place in the organization at the time of project launch. Consequently, there may need to be some development or support to ensure effectiveness in those roles. However, if stakeholders want to include the Decision Areas for the Champion role, they can use the Write-in feature described in the Project Pre-Check processes.

The following material has been adapted from a Managing Organizational Change program developed and conducted by ODR (now Conner Partners) and a number of other management of change sources. The attributes covered below can be used on a Decision Area basis to arrive at an overall comfort level for that Decision Area. For example, when evaluating the Decision Area, Sponsor Commitment, consider all the attributes listed but arrive at an overall judgment. However, if the Decision Areas method reveals significant concerns about one or more of the stakeholder capability Decision Areas, or if a more comprehensive review is desired, the attributes can also be included in detail questionnaires that seek a response on each item. This can lead to a more precise understanding of the issues and possible remedies.

Roles and Responsibilities

The success of a change initiative demands that all stakeholders are actively involved and in agreement with the decisions made on the project. It's also vital that all stakeholders understand their primary role and the primary role of every other stakeholder on the project, either sponsor, change agent, target or champion.

Sponsor

The sponsor is the project COO—the Chief Operating Officer. He or she is accountable for the overall success of the venture—return on investment, intangible benefits, costs, risks, timing, strategic and operating impact, client satisfaction, cultural impacts and all the other aspects of an organization's universe that can be altered by a major business or technology change. The sponsor's decision making focus is on the 5 W's: Who, When, What, Where and Why

Unlike a corporate COO however, all the other stakeholders do not necessarily report to the sponsor outside of the project structure. Therefore the sponsor may not hold absolute power on all decisions. For example, the sponsor of a new product launch may be the marketing VP. One of the other stakeholders in a target role could be the VP of administration. If there is a dispute or disagreement between the two, the sponsor can't simply overrule the VP of administration. The issue needs to be escalated to their common superior within the organization's hierarchy.

That's way I always like to show a project's organization as a round table, with links from each stakeholder back to their respective organizations, as shown in Part III. It makes it very clear that collaboration needs to be the standard operation principle, not command and control.

Target

A target stakeholder is a manager of an organization whose people, processes and products or tools will have to change for the project to be successful. The first priority for a target is to ensure the effective ongoing operation of the people, processes and practices for which he or she is accountable. Of course a target needs to support the planned change, but in a manner that ensures the ongoing effectiveness and integrity of the target's organizational responsibilities.

Where a project's focus includes changes that will be detrimental to a target's operations, the target should seek to change the project's directions and plans by collaborating with the sponsor and other stakeholders. If the issue can't be resolved amicably, it should be escalated.

Like the sponsor, a target's decision making focus is on the 5 W's, but with a focus on his or her operational accountabilities.

Decision Framework

Change Agent

On major business and technology change initiatives, the change agent role is most often filled by a project manager or project director. On smaller changes, the role can be, and often is, filled by a line manager of senior staff member. Either arrangement can be successful as long as the individual in the role has the requisite skills and capabilities for the job at hand and the support of the other stakeholders on the project.

Unlike the sponsor and target roles, the change agent's primary focus is on only one question: How? That is, how to implement the project successfully as imagined by the other stakeholders. Whereas the sponsor and target are most concerned with what it will look like after the change has been implemented, the change agent is most concerned with how to get there. That should provide a healthy tension to the project.

Where the sponsor and targets are visioning certain features, functions and capabilities that will lead to their desired states, the project manager is pointing out the cost, schedule and quality implications, the dependency and resource issues, the additional risks, etc., and offering various approaches and alternatives that can satisfy the expectations. The major challenge for the change agent is getting agreement from the other stakeholders on all the factors that will influence project success. Just like herding cats!

Champion

Champions represent a constituency that is critical to project success, for example a senior, well respected sales person working with sales staff to make the change. Their role is to help those affected by the planned change make the transition from the old ways to the new ways as quickly and effectively as possible.

Champions need to understand and believe passionately in the planned change. They need to be senior enough to be able to influence stakeholder group decisions. They need to be knowledgeable enough to understand the impact of the planned change on their constituents and identify and promote changes that will enhance implementation effectiveness and project success. They need strong communication and leadership skills and have sufficient respect among their constituents to be able to sway beliefs, opinions and behaviours.

Most major business and technology changes will affect multiple organizations and their staff. Is a champion required for each one? Not necessarily. Appoint a champion only for the groups that are vital to project success, have a big change ahead of them and have an individual in their midst that has champion qualities. Better no champion than a bad champion!

Target Skill Requirements

Every major change requires human beings to acquire new knowledge and skills and demonstrate new or changed behaviours. Identification of the specific target audiences affected, the required knowledge, skills and behaviours for each, and the planning to create those new abilities needs to begin at inception and continue through to the successful completion of the change.

Environment Domain

According to material developed by Robert S. Kaplan and David P. Norton, only 10% of organizations execute their strategy. Barriers to strategy execution include:

o   Vision barrier—only 5% of the workforce understands the strategy
o   People barrier—only 25% of managers have incentives linked to strategy
o   Management barrier—85% of executive teams spend less than 1 hour per month discussing strategy
o   Resource barrier—50% of organizations don't link budgets to strategy.

It's not surprising then, that very few organizations succeed in creating and managing a comprehensive view of all dimensions of the current infrastructure, or identify a target future state for all of those same dimensions, let alone manage a business plan that clearly delineates and positions all initiatives.

However, some degree of infrastructure management is essential to ensure corporate assets are delivering optimum value. Similarly, future visioning and business planning are necessary to ensure the future success of the enterprise.

Projects whose relationship to and impact on the current business environment, the target environment and the business plan are well understood have a greater chance of success, cost less, have less risk and deliver greater value to the enterprise.

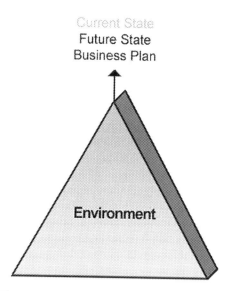

Figure 15—Environment Domain Factors

The Environment Domain addresses key dimensions of the current state enterprise infrastructure, the future state target architecture and the business plan that will move the enterprise from the current state to the desired future state. *FastPath* asks stakeholders for focus on the future state and business plan.

Business Plan Factor

An organization's Business Plan, which encompasses key facets of an organization's reason for being, hopes, aspirations and directions, provides the context for any business or technology change. The relationship of a planned change to each of these facets should be clearly understood by all those involved in and affected by the change to ensure consistency of direction and ongoing success. Any dichotomies between the stated directions and beliefs and the impact of a planned change need to be addressed as part of the change effort.

Business Plan Impact

Do you know how the project you're working on supports the organization's mission and vision? Have you considered how the planned change aligns with or is at odds with the organization's culture and core values? Do you know how it contributes to enterprise strategies and priorities? If you don't, you're operating without some critical information needed to ensure project success. Here's some guidance on what you need to know about the relationship between your project and your organization's business plan and how it can help you and your stakeholders improve project performance.

The Decision Framework Business Plan Factor includes eight Decision Areas: mission, vision, core values, strategies, enterprise priorities, dependencies, programs and timing. In *FASTPATH,* the focus is on a hybrid Decision Area—Business Plan Impact—which includes strategies and enterprise priorities.

Strategies

A strategy shows how to achieve a vision. It's a game plan. There should be a very clear relationship between the planned change and one or more strategies. The tie-in should be documented and communicated to all involved in the planned change.

Enterprise Priorities

Enterprise priorities delineate the specific initiatives that will be done to support the strategies, the sequence of conduct, the dependencies and the levels of funding. The fit of the planned change within the enterprise priorities should be fully disclosed to all involved.

The following case shows what can happen when priorities and dependencies aren't clearly understood.

Sometimes Agile Isn't

The Situation

This financial services organization had attempted to replace the system used for the production of statements for its clients and their employees on three occasions over the last ten years. The statement runs produced over one million statements each quarter consuming annual processing costs of about $200,000. The applications were a mish mash of legacy technology and code that was very difficult to change and problematic when it came to ensuring the necessary quality levels. Previous statement projects usually took up to twelve months to implement and cost hundreds of thousands of dollars.

The Goal

The business wanted to replace these applications with a new system that would be much easier and faster to change and at the same time provide support for a variety of ad hoc queries for administrative and sales staffs and their clients. The new solution had to provide the ability to replicate exactly previously produced client and employee statements at any point in the future. In addition, annual production costs shouldn't exceed $250,000.

The Project

The statement project was commissioned to tackle both the production of statements and support ad hoc queries through the creation of a data warehouse.

The VP of the business unit was the project's sponsor. Two project managers were appointed, one from the business and one from IT. Selection of new technology to support the data warehouse and the ad hoc queries was seen as a priority so a technology assessment stream was launched to expedite that process. In addition, because of the difficulties that had been experienced in the last three attempts to solve this problem, senior IT staff proposed the use of agile techniques to allow phased delivery that could be incrementally built upon to achieve the final state. The IT PM, a recent hire, was selected in part because he had previous agile experience.

The new technology was selected, installed and vendor staff was brought in to help train the in house staff and build the application. Concurrently agile training was selected and conducted under the tutelage of a contract agile leader. In short order the project team was up to 18 IT and contract staff and burning through $200,000 per month. A data warehouse proof of concept was launched in the fourth month to bring all the disparate elements together—production statement, ad hoc queries, new technology and agile.

The Results

The ad hoc aspect suffered because the business hadn't really thought through what it was they wanted to do. The agile implementation suffered because the team wasn't able to identify and tackle implementable pieces. Much of the work reverted to traditional development practices the majority of the staff were used to. Finally, a production statement pilot revealed that annual processing costs would be in excess of $ 2 million using the new technology versus $200,000 for the existing system. After another two months of analysis and head scratching, the project was cancelled with over $ 2 million in staff and contractor costs down the drain.

How a Great PM Would Have Helped

This project is a sad example of how not to implement change effectively. A great PM would have done a number of things differently. For example:

- Look at the reasons for the failure of the previous three attempts to deliver an improved environment to support production statements and ad hoc queries. In all three cases, the project wasn't really a top business priority. Business support eroded as the demand for more decisions escalated. Getting clear direction on the Project Pre-Check Change Domain Decision Areas, like burning platform, opportunities, goals, worth, requirements, benefits, assumptions, etc. would have revealed the clarity of the business vision as well as the commitment of the stakeholders.

Decision Framework

- One of the challenges this project experienced was getting the required business resource. Apparently the business had reorganized about the time the project launched and the reorganization consumed the business management time that was required for the project to progress. Déjà vu all over again!
- This project was actually four different changes grouped together: product statements, a data warehouse with ad hoc query capability, new technology and agile. One of the Decision Areas in the Change Domain mentioned above is assumptions. If the PM had explored the assumptions around each one and the four together, I expect a very different structure and approach to the problem would have been used.
- One of the sources of ongoing friction between the business and IT was the use of the data warehouse for the generation of production statements. IT believed the profiles of the two functions were sufficiently different to warrant unique solutions. The business disagreed. Had the PM considered the Volumes, Mix & Peaks Decision Area in the Project Pre-Check Change Domain, he would have discovered vastly different profiles, well known for the production statements, yet to be determined for the ad hoc queries. Had he had committed business stakeholders, the accumulated evidence would have helped them see the light.
- There was no risk analysis or plan, an interesting oversight for a project that had failed three times previously. A risk plan would have considered the challenges posed by running the four changes together, would have looked at the impact of the business reorganization on project resourcing and would have addressed the challenges associated with implementing and operating a new technology component. The resulting mitigation strategies may have helped avoid project failure and could have guided the PM to a different approach.

Every time I run into a project like this, I get angry. Please get your priorities straight! It's so easy to use a framework like Project Pre-Check to make sure the PM and other stakeholders address factors that have proven essential to project success. It's an almost painless exercise, it takes little time and it cements the stakeholders' commitment to the endeavor. If you don't want to use Project Pre-Check, build your own checklist of vital best practices, use it on every project and shape it to reflect your own learnings. Please! Your stakeholders will thank you.

There is a silver lining to this project. The PM facilitated a project post mortem and developed a lessons learned log covering many of the points above. The internal team was moved largely intact to a project in another business unit where they have been able to apply their learnings and leverage the benefits of agile techniques.

Future State Factor

While the word "environment" is most often associated these days with "being green", the meaning in Project Pre-Check *FastPath* relates to the world that the planned change will operate in, and all of the aspects of that world that the planned change will disrupt, improve, add to or eliminate. Here are some factors to consider to give you and the other stakeholders a project world view.

To arrive at your project world view, consider the impact a change has on the following areas as early as possible in the project life cycle and update that view as the project progresses and as new insights are gained. Consider the affect the planned change has on the current state (what exists today that may change as a result of the planned change) as well as the future state (what is planned for the future that may change as a result of the planned change) and the inter-relationships between them. These last two views are not always easy to gain but are often very revealing and beneficial for the project.

Keep in mind that only a high level view is needed initially. That gives stakeholders the information they need when molding the change into the optimum shape. It also provides a framework for the follow-on work that will be done as part of the project. The factors you need to consider are:

External Relationships

External relationships identify and define new operating relationships or existing relationships altered by the change, and include:

- o customers
- o markets
- o suppliers
- o channels
- o regulators
- o partners
- o distributors
- o industry organizations
- o competitors
- o etc.

Products and Services

Products and services include any new offerings that the enterprise provides to clients or changes to existing products and services required by the change.

Processes and Functions

Processes and functions include new processes and supporting operations that will require changes as part of the planned project. Processes are typically identified and defined from the client in. Functions relate more to internal operations such as accounting, payroll, regulatory and governance operations, etc.

Interfaces

Interfaces relate to all unique interfaces—human and machine—with the external world and within the operations of the enterprise that will be required or affected by the change. This includes reports, screens, web sites, commercials, technology feeds or bridges between internal or external systems, regularly scheduled seminars and conferences, etc.

Information

The Information factor includes any new needs or changes to the existing collection of facts or data that support and enable the core business operations. This can include new requirements or changes to the information that is needed about clients, products, partners, suppliers, regulatory agencies, competitors, etc.

Technology

Technology encompasses new capabilities and changes to existing hardware, software, communications and associated services needed to support the enterprise including:

- o   hardware assemblies and components
- o   operating and application software

- o communications capabilities and services
- o technology capabilities and services
- o application services
- o vendors
- o etc.

Resources and Facilities

The Resources and Facilities Decision Area includes new requirements and changes to the existing human, financial and other resources and facilities that are required to support the organization including:

- o people by job level, skill set,
- o floor space and related facilities
- o physical plant (owned, leased and rented)
- o services (security, cleaning, audio/visual, video conference, etc.)

Organization

Organization includes new organizational units or structures and changes to current organization structures, accountabilities and relationships that are required to achieve operational goals and objectives.

Remember to ask the question, "what impact will the change have on the current state and the planned future state of these environmental areas?". Make sure that all stakeholders agree with the answers you get.

Assets Domain

The Assets Domain includes the people, processes, products and tools that can be leveraged to support a planned project or may require changes for the project to succeed. In essence, the Assets Domain is the supply cabinet to the project world.

If the people required for the planned change can be acquired and deployed with current practices and processes, without the need for training or skill development, using existing equipment, services and physical plant, and they can use existing processes and technology to develop,

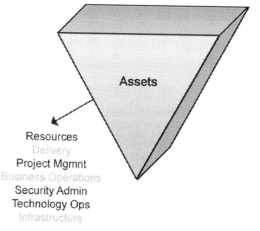

Assets

Resources
Delivery
Project Mgmnt
Business Operations
Security Admin
Technology Ops
Infrastructure

Figure 16—Assets Domain Factors

deliver and support the change, that is a significant advantage to the planned project. By re-using existing assets, the scope, costs, time and risks can be reduced, perhaps significantly.

Conversely, if new staff or skills are required, new tools or technologies are necessary or new or changed processes and practices are needed to develop, deliver and support the change, these new or changed elements should be included in the overall scope of the project. The costs, resources and time required to address the changes need to be reflected in the overall plans, costs and risks of the undertaking. This can add a significant burden to a planned change and must be addressed early in the change life cycle to avoid unpleasant surprises and negative consequences as the project progresses.

Considering the impact of a planned change on the Factors and Decision Areas in the Assets Domain can help stakeholders identify and deal effectively with those other, significant, intangible assets in a preemptive manner. There is, however, a philosophical question that needs to be considered by the stakeholders: does the success of the project require every process or practice to be exactly as needed or can some or all of the elements supporting or affected by the change be "just good enough". The answer can have a huge impact on the time, effort, costs and risks involved in delivering an effective solution.

The following Factors are addressed in the Assets Domain in *FastPath*: Resources, Project Management, Security Administration and Technology Operations. In *FastPath*, the Decision Areas under the Delivery, Business Operations and Infrastructure Factors have been excluded.

Resources Factor

The Resources Factor encompasses the following sources in terms of being able to support the planned change:

- o   internal staff within the business and IT organizations
- o   contract staff
- o   client participation
- o   vendor and partner support
- o   internal and external audit services
- o   regulatory agencies
- o   technology, business, physical plant and building services

The Resources Factor includes seven Decision Areas including skills and capacity, skill development, resource procurement, contract management, team formation, performance management and succession planning. In *FastPath*, only the skills and capacity Decision Area is included.

Skill & Capacity

For any change to be successful there must be sufficient numbers of people with the necessary knowledge and skills who are or will be available when needed for the requisite amount of time and in the required locations.

Decision Framework

Additional considerations need to include:

o the type of resources (executive, line, staff, business, system, technical, professional, etc.),

o the sources (internal, partners, customers, contract, recruiting, etc.),

o the costs of the sourcing alternatives

o the priorities of, impact on and tradeoffs associated with other competing assignments.

It is unusual in these days of rapid business and technology change to encounter a situation where all necessary skills and capacities are readily at hand. In most cases, a planned change requires some new knowledge or skills or additional capacity to be successful.

An existing, effective skill planning and development process and the requisite facilities can be used to build and deliver new or revised training programs quickly and effectively to meet the demands of the planned project and the diverse target groups affected by the change.

Project Management Factor

The project management Factor within the Assets Domain encompasses the project management processes and practices that are available within the enterprise and can be leveraged to manage the planned change. Stakeholders need to consider the existing, standard practices within the organization that have been proven over a number of projects and whether they can be applied effectively to the project in question.

The Project Management Factor includes the following Decision Areas: organization, estimating, project tracking and reporting, requirements management, risk management, issue management, change control, defect tracking and gating. In *FastPath*, four of the ten Decision Areas are included; processes and tools, risk management, issue management and change control.

Project Management Process

A tried and true project management process and the supporting tools and techniques are necessary prerequisites to help an organization launch projects quickly, plan and control progress effectively and deliver solutions that are appropriate to the need. A process should specify the steps that need to be considered, provide a variety of profiles for different sizes and kinds of projects, include specific practices that can be leveraged by the project team and identify tools and techniques that can enhance team effectiveness.

Stakeholders should look for evidence that a standard project management process is in place, has been used successfully with similar projects and is familiar to project management and staff.

Risk Management

The Software Engineering Institute (SEI) uses the Webster's definition of risk: risk is the possibility of suffering loss.

The SEI presents a six step process for continuous risk management:

- o Identify: Search for and locate risks before they become problems.
- o Analyze: Transform risk data into decision-making information. Evaluate impact, probability, and timeframe, classify risks, and prioritize risks.
- o Plan: Translate risk information into decisions and mitigating actions (both present and future) and implement those actions.
- o Track: Monitor risk indicators and mitigation actions.
- o Control: Correct for deviations from the risk mitigation plans.
- o Communicate: Provide information and feedback internal and external to the project on the risk activities, current risks, and emerging risks.

Whether an in place process mirrors the SEI view or reinforces an organization's unique needs and experiences, an established risk management process is an extremely valuable asset. Without an existing process to leverage, significant time, effort and energy may have to be devoted to defining and implementing appropriate means of identifying, quantifying, mitigating and managing risk.

Issue Management

Issue management is the process used to deal with problems that may jeopardize the project's ability to deliver to stakeholder expectations. Issues may relate to any problem perceived to exist by an interested party and can include concerns re functionality, quality, resourcing, skill sets, timing, benefits, costs, value, markets, competitors, customers, partners, etc.

To resolve a project problem successfully, everyone involved needs to understand that they are a part of the process. One of their primary responsibilities is to raise issues when they see them. Raising an issue should not be viewed as a negative event but as a proactive way to bring a problem to the surface so that the stakeholders and the project team can apply appropriate resources, find alternative solutions, and implement a resolution.

Having a defined, well-used issue management process enables issues to be exposed and a solution implemented quickly and effectively. Stakeholders can reinforce the use of the issue management process by responding quickly and positively to issues raised and by recognizing and rewarding staff who identify and submit issues for resolution.

Change Control

A change control process is essential for managing requested changes to a project, including changes to project scope, deliverables, finances, milestone or resources. It structures and directs stakeholders' actions toward reaching an informed and timely decision based on benefits, costs and contribution to project success.

In addition, the change control mechanism must become operational immediately on submission and acceptance of the initial estimates. That way, any changes in scope or intent, or changes to the estimates themselves, can be vetted through the change control process and approved by stakeholders.

"The one unchangeable certainty is that nothing is certain or unchangeable."

John F. Kennedy

Security Administration Factor

The International Standards Organization has released ISO/IEC 17799:2005, an International Standard "Code of practice for Information Security Management". A summary of the standard, available online, states: "Information is an asset that, like other important business assets, is essential to an organization's business and consequently needs to be suitably protected. This is especially important in the increasingly interconnected business environment. As a result of this increasing interconnectivity, information is now exposed to a growing number and a wider variety of threats and vulnerabilities."[27]

The 2005 version of the standard contains eleven main sections, all of which are appropriate for development or assessing the appropriateness of an overall security program. For the purposes of Project Pre-Check however, where the emphasis is on assessing the value to the planned change or the impact of the planned change on an enterprise asset, it's reasonable to focus on the value from and impact on the following four aspects of an overall security program: security policy, physical security, information security and human resources security.

$11.9 billion—the estimated cost of viruses, worms and Trojan horse programs to U.S. organizations

Source: FBI survey of 2066 organizations

Security Infrastructure

Security Policy

According to the ISO/IEC standard, the security policy provides management direction and support for information security. It should state the level of management commitment and provide guidance regarding the policy's implementation. It should include a brief explanation of the security policies, principles, standards and compliance requirements of particular importance to the organization, for example:

o   Compliance with legislative and contractual requirements
o   Security education requirements
o   Prevention and detection of viruses and other malicious software

Decision Framework

- o Business continuity management
- o Consequences of security policy violations

It should also provide a definition of general and specific responsibilities for information security management, including reporting security incidents.

Stakeholders need to understand the value that an in place security policy can provide to the planned change. Any revisions needed to the policy as a result of the planned change should be included in the project scope.

Physical Security

The physical security arrangements and practices should prevent unauthorized access, damage and interference to business premises and information and prevent loss, damage or compromise of assets and interruption to business activities. They address the following considerations and practices:

- o Housing of critical/sensitive business information processing facilities in secure areas
- o Physically protecting assets from unauthorized access or damage or interference
- o Articulating and administering clear desk and clear screen practices
- o Protecting equipment physically from security threats and environmental hazards
- o Reducing risk of unauthorized access to data, to protect against loss or damage
- o Equipment siting and disposal
- o Safeguarding of special controls to e.g. electrical supplies

Stakeholders should be aware of the value existing physical security arrangements deliver to the planned change. Any changes needed to the physical security practices as a result of the planned change should be included in the project scope.

Information Security

The information security processes and practices should address protection of information and information processing facilities from disclosure, modification or theft by unauthorized persons and provide controls to minimize loss or damage. They deal with the following considerations and practices:

- o Accountability for assets
- o Information classification practice to ensure that information assets receive an appropriate level of protection
- o Classification of information to indicate need, priorities, degree of protection, degrees of sensitivity, criticality
- o Defining appropriate set of protection levels, communicate need for special handing measures.

Stakeholders should understand the value that existing information security practices provide to the planned change. In addition, any changes required in the information security practices as a result of the planned change should be included in the project scope.

Human Resources Security

Human Resources Security focuses on reducing the risks of human error, theft, fraud or misuse of facilities. It ensures that users are aware of information security threats and concerns, and are equipped to support organizational security policy in the course of their normal work. Human Resource Security covers the following considerations and practices:

- o Screening potential recruits, contractors, vendors and partners
- o Addressing security responsibilities at the recruitment stage, including contracts and monitoring during employment or engagement
- o Acquiring and maintaining confidentiality agreements
- o Training users in security procedures, correct use of information processing facilities to minimize security risks
- o Reporting security incidents and the procedures for security breach, threat, weakness or malfunction
- o Establishing formal disciplinary process to deal with violators

Decision Framework

Stakeholders need to consider the value the current human resources security practices deliver to the planned change. Any changes required in the human resources security practices as a result of the planned change should be included in the project scope.

Technology Operations Factor

The Technology Operations Factor encompasses the in-place technology management processes that will be required to support the planned change during its development and on and after implementation. There are a number of in-depth industry practices that view technology services with a much more granular perspective including the IT Infrastructure Library (ITIL), from the Office of Government Commerce (OGC)[28] and the Microsoft Operations Framework (MOF).[29]

For the purposes of Project Pre-Check however, the breadth and depth from these sources and others has been consolidated into six Decision Areas that can be easily understood and assessed by all manner of stakeholders relative to a planned change. It includes the service desk, change management, system administration, output management, technology installation and service level management.

Stakeholders need to assess the availability of these processes and the suitability to the planned change. If the necessary processes do not exist, are not adequate for the needs of the planned change or must be revised to support the planned change, then the gaps need to be reflected in the work plan, costs and schedule for the planned project.

Service Desk

The Service desk provides support to the user community for issues and problems associated with the use of application and technology services. It is typically responsible for dealing with user queries and problems and for overseeing the restoration of normal service.

The three main focuses of the Service Desk are incident control, resolution and communication. Services provided can range from simple call logging functions and escalation to incident management and status communication to problem resolution.

It's important to assess the impact of a change on Service Desk operations and capabilities to ensure that support is readily available to service users. Factors to consider include:

- o Changes in client base
- o New or changing clients
- o Knowledge and skill levels
- o New or changing locations
- o Language and cultural support requirements
- o Changes in potential call volume
- o Peak and average profiles
- o Hours of coverage
- o Changes in the technology supported
- o Changes in the applications supported
- o Changes in service level agreements
- o New metrics or measurement requirements
- o Organizational or relationship changes affecting escalation protocols

Change Management

The Change Management processes and practices ensure that standard procedures are used for accepting, recording, prioritizing, planning, testing and implementing changes, with minimal impact on services. It assesses the impact on and proves appropriate ongoing operability relative to established service levels, security policies, configurations, capacities, performance, availability and service desk operations.

Stakeholders should be cognizant of the impacts of the planned change on the ability of the change management process to operate effectively and include any new or additional requirements in the project scope. Factors to consider include:

- o The change request mechanism
- o Change classification and prioritization
- o Change authorization and approval process
- o Change implementation
- o Change review process

System Administration

System administration is responsible for day-to-day tasks of keeping enterprise systems running, for assessing the impact of planned releases, monitoring the health of IT services and acting when necessary to maintain compliance.

Stakeholders should consider the following dimensions of the system administration Decision Area to understand the potential impact of the planned change and identify any incremental effort required in the project scope:

- o Directory services—allow users and applications to find network resources such as users, servers, applications, tools, services, and other information
- o Job scheduling services—handle the sequencing of various computing tasks (printing, database, backups, and others) for optimal use of computing and network resources
- o Network administration—operates basic network services on a day-to-day basis
- o Security administration—deals with the routine application of security policies and practices to maintain a secure operating environment
- o Service monitoring and control—observes the day-to-day health of the operating environment, monitors and resolves incidents and provides alerts to interested parties
- o Storage management—is the set of practices dedicated to safe, secure storage of data, effective backup-and-restore policies, and efficient use of storage resources

Project Domain

The Change and Environment Domains identify the Factors and Decision Areas that needed to be addressed to understand and fully articulate the planned change. The Assets Domain identifies Factors and Decision Areas representing processes and practices that may be affected by the planned change or that could be used, as is or with modifications, to support a project.

The Project Domain, on the other hand, addresses the specific project decisions that need to be made to shape and deliver changes that achieve the organizations goals. These

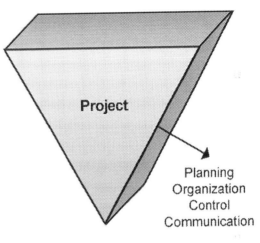

Figure 17—Project Domain Factors

decisions deal with how to deliver to the stakeholder expectations articulated through the decisions made in the Change and Environment Domains. In other words, the Factors and Decision Areas in the Project Domain deal with "how" to deliver the "what" established in the other Domains.

The Project Domain includes four Factors: Planning, Organization, Control and Communication.

Planning Factor

The Planning Factor within the Project Domain requires the application of relevant best practices addressed in the Assets Domain to deal with the unique needs of the planned change as articulated through the Decision Areas in the Change and Environment Domains.

Decision Framework

It covers the following Decision Areas: business alternatives, technology alternatives, release plans, cost estimates, benefit plan, quality plan, resources and facilities plan, contracts plan, communication plan and risk plan.

Business and Technology Alternatives

Fishbein's Conclusion: The tire is only flat on the bottom.

Stakeholders have a responsibility to ensure that a rational and thorough identification and evaluation of business alternatives is completed before selecting the preferred course of action. Often, the formal assessment exercise can yield significant opportunities to deliver the desired results in less time, at less cost and risk and with greater competitive advantage. The Decision Areas in the Change Domain provide the framework for a thorough evaluation of the alternatives.

As with business alternatives, a formal identification and evaluation of technology alternatives is essential. The Decision Areas in the Change, Environment and Assets Domains provide the foundation for the evaluation criteria that can be used to review each alternative and arrive at a decision.

Release Plan

"Little by little does the trick."

Aesop

One of the most frequently espoused best practices is the "chunking" of project elements into smaller, implementable packages called releases. The advantages include reduced risk, an acceleration of the benefit stream and greater flexibility to react to changes in needs, priorities and the marketplace. In fact, there is compelling evidence that smaller implementations are almost always more successful.

The release plans for a given change should specify, explicitly, what will be included in each release, when it will be delivered and to whom. The plans should also demonstrate compliance or divergence with the Change and Environment Domains' Decision Areas.

The Release plan should include the following information:

- o Release alternatives including prototyping, time boxing, phasing, staging and partitioning options.
- o The recommended content and structure of all releases and the rationale for the recommendations in terms of responsiveness cost and benefit effectiveness, quality and risk.
- o Mapping of each release and the overall plan to Change and Environment Domain Decision Areas to demonstrate compliance with those decisions or identify recommended variances.
- o A milestone plan for the initial release.
- o The overall timeframe for each release and for the completion of the change.
- o Assumptions and constraints.

The change agent should recognize that the release plans, while driven by the business phasing and staging priorities, also need to encompass all other factors that may influence the viability and effectiveness of the various release options and the final plan.

Here's an example of how insightful release planning can make the difference between success and something much less desirable.

Avoiding New Technology Risks

The Situation

This U.S based financial services organization managed 401(k) plans for employees of small to medium size businesses. A 401(k) retirement savings plan allows a worker to save for retirement and have the savings invested while deferring current income taxes on the saved money and earnings until withdrawal.

The company's product offerings allowed participating client employees to invest in a wide variety of popular mutual fund products. The client employees would call the company's service staff to get quotes on funds and to place orders for new investments or changes to existing ones. The company had a web site that offered similar services but found that the volume of phone calls continued to increase with the growth in their business. It was posing an ever increasing cost burden on the organization.

Decision Framework

While interactive voice response (IVR) technology is common today, in the early 21st century it was an emerging technology. However, because it held such promise to address their challenge, the company decided to take the plunge.

The Goal

The VP, Customer Service, the sponsor of the initiative, after discussions with the CIO and senior technical staff, decided to wade into the then largely uncharted IVR waters. She launched a project to leverage the new technology to reduce the current and growing costs of their phone services.

Her goal was to deliver an IVR system that would support the quote and trading functions for client employees across the country and have it fully operational in one year. She was looking for a 50% rate of conversion to IVR.

The Project

The IT organization appointed a seasoned project manager. Because the company had no experience in the IVR arena, he knew trying to estimate the costs of the project would have been a fool's errand. They talked to a number of vendors and early users to know that an effective solution was possible. But the costs quoted varied widely. So he worked with the VP to figure out what she could afford by looking at the expected reduction in current costs and slowing the cost growth rate while offering the same or improved service levels. Her target—$1.2 million with a 70% IVR usage rate. That would give her a 2 year payback.

The project manager, working with an assigned business manager, the $1.2 million cost target and two of the most likely technology vendors, developed a plan of attack that included the following key elements:

- Identify and engage key stakeholders
- Get stakeholder agreement on the dimensions of the change. That included clarifying the opportunity for the company beyond this particular application, the specific goals, requirements, benefits, locations affected, desired or required target dates, anticipated volumes and phasing and staging opportunities from a business perspective.
- Get stakeholder agreement on the quality expectations, including factors like security, authorization, ease of use, continuity, scalability, localization and audit trail.

- Also, with the stakeholders, firm up the economic justification including the overall economic impact (business, IT, clients, others), competitive advantage, strategic fit, competitive risk and project risk.

With a fully developed picture of the project, its impact on the organization, and buy-in from the stakeholders, the PM assembled a small team including an experienced and respected staff member from the Customer Service organization, a business analyst, a programmer analyst and a senior technical analyst from the IT Operations group. Together they developed an RFP that went out to the two IVR vendors who had helped in the planning plus two additional vendors that showed promise based on feedback on the systems they had delivered.

When the RFP responses came back, the content was vetted, rated, ranked and their proposals assessed against the $1.2 million cost target. One company stood out (one of the two involved in the planning effort) and was approached to deliver the required solution. During the contract negotiations that followed, it was also decided to rely on the vendor to do the application development and maintenance rather than train up internal resources to develop the applications and maintain them after implementation. This would avoid the risks involved in having a critical technology service being supported by a too small internal talent pool.

Once the vendor was selected and on side, the project proceeded with a five stage plan:

- Develop an initial prototype focusing on the quotation function to test the integration of the technology into the company's infrastructure and the required application interfaces and get stakeholder reaction to the speech recognition structure and sound.
- Implement the full quote capability in one region of the country to ensure the system's ability to support the local accents, assess the willingness of their clients to use the new service and refine as needed to address any issues. In this first implementation, clients would be given a choice of using IVR or speaking to a Customer Service consultant as before.
- Roll out the quote capability across the country, region by region, assessing support for local accents and degree of use in each region and refine as necessary. The choice to use IVR or speak to a person would continue to be offered.
- Develop and implement the full IVR trading capability and implement in one selected region as before to gauge effectiveness and appeal and revise as necessary.
- Roll out the full trading solution across the country, one region at a time, refining as necessary.

Decision Framework

The Results

The project was a terrific success. It was fully deployed across the country in 11 months at a cost of $1.1 million with IVR utilization at 75% and rising. Customer feedback was very positive, especially because of the extended 24 hour phone service window using IVR versus the old 12 hour service period.

The staged rollout allowed the vendor to tweak the application to be more sensitive to local accents and adapt the structure and content of the menus and messages to improve responsiveness, based on feedback. It also allowed the organization to target clients and employees in the region being implemented to introduce the new IVR service, encourage client employees to use it and explain how to address problems and provide feedback.

The staged rollout gave the business the ability to redeploy the Customer Service staff gradually over the 6 month rollout period, reducing the anxiety that is a normal part of business and technology change. Also, it gave the IT organization the time needed to integrate the new IVR technology into their infrastructure and operations and refine the application interfaces to improve IVR responsiveness. And, all of the stakeholders were thrilled with the outcome. It was a project well done!

How a Great PM Succeeded

This project manager did a number of things right:

- He leveraged Project Pre-Check's three building blocks (even though he had never heard of Project Pre-Check).
- He identified and engaged the key stakeholders, including the vendor and the clients.
- He took them through a structured process that enabled them to stay interested and involved, allowed them to provide their expertise to the project from inception through completion and confirmed their agreement with decisions made at each stage of the project.
- Finally, he leveraged best practices (the equivalent of Project Pre-Check's Decision Framework) to ensure the factors that should be addressed to ensure success were, in fact, addressed.
- The staged, iterative approach he took to implementation effectively avoided the risks of a big bang approach and allowed the vendor and project team to adjust the applications, the environment and the internal practices based on real world experience.

- His work with the sponsor to help her figure out how much she could afford to spend on the project to achieve her financial and service goals (it's call 'Worth' in Project Pre-Check's Decision Framework) was a key success factor. It was a meaningful figure that drove all the project decision making from scoping, RFP and vendor selection, through implementation. It enabled the other stakeholders and project team members to make calls against a meaningful number they all understood.
- He managed to avoid the usual potholes and pitfalls usually associated with trying to leverage new technologies to deliver business value.

Cost Estimates

The cost estimates for a project should be driven by the Release Plan. They need to include all one time and ongoing costs and incremental capital and operating expenses for external and internal resources and facilities.

The change agent should ensure that stakeholders are thoroughly familiar with the estimates for a given project and be comfortable with the methods chosen and the practices applied to arrive at the estimates before committing their scarce resources to the effort.

By comparing estimated to actual results early and often, understanding the reasons for any variances and correcting or adjusting the underlying assumptions on which the estimates are based, the estimates will evolve to provide a meaningful and powerful management tool.

Quality Plan

A quality plan describes the steps that will be taken and the resources required to demonstrate that all project expectations have been achieved. A quality plan typically includes the following elements:

- o The quality targets, based on the Decision Areas in the Change and Environment Domains.
- o The standard practices and processes that will be applied and any other practices required by the organization, industry or jurisdiction.
- o The components, services, facilities and deliverables that need to be tested.
- o The methods and mechanisms that will be used to assess the quality of each element.

Decision Framework

- o The timing of each testing exercise and the dependencies between each test.
- o Accountabilities for each test and the delivery of results that achieve the established targets.
- o Assumptions and constraints.

Resources and Facilities Plan

The Resource and Facilities Plan identifies the resources, services and facilities required by the project. It also specifies the various dependencies, the timing required and the procurement processes and practices that will be executed. It will reference the pertinent Asset Domain Decision Areas addressed in the Resources and Project Management Factors.

The Resource and Facilities Plan ensures that internal and external human resources, technologies, services and facilities are available when needed to support the release and quality plans. The plan needs to encompass business and technology, management and staff, customer, partner, supplier and any other resources and facilities on which the success of the change is dependent.

The plan should include the following information:

- o The various skills and capabilities required and the numbers of each over time.
- o Anticipated skill and capacity gaps and the processes and services that will be used to address the needs.
- o The planned utilization rates.
- o The amount and type of technology, services and facilities required, both internally and externally, the timing for delivery and the processes that will be used to deliver.
- o The planned time frames for delivery and use.
- o Individual responsible for the acquisition and management of the resources.
- o Any assumptions and constraints relating to the resources and facilities plan.

Communication Plan

The communication plan for a project is concerned with getting the right information to the right parties at the right time. It needs to address the needs of all parties who will be involved with and affected by the planned change. It should reference decisions reached regarding the

Project Tracking and Reporting Decision Area under the Project Management Factor and the Corporate Tracking and Reporting Decision Area under the Business Operations Factor, both within the Asset Domain.

The communication plan can include unique materials for each target group, timed to maximize the value of the communication and can include any or all of the following approaches:

o   An ongoing series of written information such as project plan documentation, project schedules and status updates targeted to the needs of the various audiences
o   Briefing sessions
o   Issues/change request tracking reports,
o   Information gathering sessions
o   Web sites, blogs, RSS feeds and other social media tools with current status information
o   Town halls
o   Questionnaires re progress, satisfaction, issues and concerns
o   Project wikis to encourage multi-way communications
o   Contests and promotions

The following factors should be taken into consideration when developing the communication mechanisms for a project:

o   Consider the organization culture and political environment (people's expectations, agendas and sacred cows, risk aversion, etc.) and ensure these are reflected in the project's communications plan. The needs, concerns, agendas and culture of each stakeholder group must be consciously evaluated and the format, structure and sharing of project information should be tailored to their specific requirements.
o   Fit the communication plan to the project's impact. All projects, whether large or small, need communication plans. More than project complexity and duration, project impact (what it promises to deliver) will most likely drive the size of the communication plan.
o   Facilitate and promote understanding through multi-way communication (be inclusive and open). Resolving conflicts by listening, facilitating and focusing on goals and not people will go a long way to promoting understanding and cooperation among all project stakeholders. Maintaining an open dialogue will create an environment of continuous feedback. People tend to feel a part of that which they are involved in.

Decision Framework

- o Plan for a crisis communications (what will you do when the worst happens?) When things go wrong the need for continuous communication is at its most critical. When chaos and crisis are the order of the day, people's emotional need to know is at its fullest. In the midst of crisis, rumors run rampant and they usually lean toward the negative side of the scale.

The Communication Plan should include the following information:

- o Target audiences—the parties involved with and affected by the change who need information and/or need to provide feedback about change and their primary roles: sponsor, change agent, target and champion.
- o Reason—the purpose for communicating with each party or group.
- o Key messages—the information and context for each party to ensure they are able to understand and act as required. More than one message may apply to each party or group or may vary over time as the project progresses.
- o Content—what each communication will contain.
- o Frequency and duration—how often a particular communication should occur, when it should commence and for how long.
- o Communication tools—the specific communication tools and methods that will be used to convey information to all parties. The most effective and appropriate tools will vary based on the audience need.
- o Responsibilities—the individual responsible for preparation, scheduling and distribution of each communication.
- o Effectiveness measures and measurement mechanisms—to ensure the various communications are achieving the desired result.

Risk Plans

"If you take risks, you may still fail. But, if you do not take risks you will surely fail. The greatest risk of all is to do nothing."

Roberto C. Goizueta —
CEO of Coca-Cola

Risk planning and management is a key best practice for project success. The Risk Plan should reflect the decisions made concerning the Risk Management Decision Area under the Project Management Factor within the Assets Domain.

Stakeholders need to pay particular attention to the following:

- o Identifying sources of risk
- o Determining the probability of a risk impacting a project
- o Assessing impact
- o Taking concrete action to manage risk
- o Prepare contingency plans
- o Managing/monitoring risk effectively on an ongoing basis.

It's one thing to identify and assess risks and develop mitigation and contingency plans. However, for all that hard work to yield the anticipated returns, the mitigation actions and contingency plans need to be integrated into the overall project plan where appropriate dependencies can be reflected and the necessary skills and resources can be allocated in accordance with the needs of the change.

Organization Factor

"If you want to manage people effectively, help them by making sure the org chart leaves as little as possible to the imagination. It should paint a crystal-clear picture of reporting relationships and make it patently obvious who is responsible for what results."

Jack Welch, former CEO of GE

The Organization Factor within the Project Domain includes those practices and decisions that are necessary to ensure the project organization structures, roles and associated resources address the unique needs of the planned change. The change agent is accountable for insuring that stakeholders address the relevant Decision Areas to their satisfaction and communicate to all involved and affected parties.

Project Organization

The change agent needs to ensure that the structure of the organization created to direct and deliver a planned change is clearly defined and the relationships to other organizations and parties are clearly delineated. The actual organization put in place should reflect decisions made relating to the Dimensions and Stakeholders Factors in the Change Domain.

Decision Framework

Just a reminder too, that Project Pre-Check *FastPath* prefers a round table organization structure over the traditional structure as discussed in Parts II and III. The round table helps reinforce the necessity of a collaborative effort for project success.

Project Roles and Responsibilities

In addition to the organization structure, the change agent also needs to ensure that the roles and responsibilities of the stakeholders filling positions within the project organization are clearly defined and communicated to the project team and target audiences.

One useful tool for showing accountabilities is the RACI chart. RACI is an acronym formed from:

R—Responsible: has a major role to perform in the process / carries out the function.
A—Accountable: ultimately owns, develops, manages, improves.
C—Consulted: must be consulted when changes are made or new opportunities are considered.
I—Informed: must be kept informed.

Identifying which stakeholders are involved in the approval of decisions on key decision points and their level of responsibility early in the change cycle is essential for rapid and sustained progress. The following RACI chart for the *FastPath* Decision Framework suggests accountabilities by role. Because of the nature of the *FastPath* stakeholder group where collaboration and agreement to each decision are paramount, only accountabilities have been shown.

Domain	FastPath Factors	Stakeholder Role			
		Sponsor	Change Agent	Target	Champion
Change	Dimensions	A			
	Stakeholders	A			
	Investment Evaluation	A			
	Quality	A			
Environment	Future State			A	
	Business Plan			A	
Assets	Resources		A		

Domain	FastPath Factors	Stakeholder Role			
		Sponsor	Change Agent	Target	Champion
	Project Management		A		
	Security Administration		A		
	Technology Operations		A		
Project	Planning		A		
	Organization		A		
	Control		A		
	Communication		A		

Table 14—FastPath Stakeholder Accountabilities

Team Formation Plan

The Team Formation Plan defines the specific series of steps and practices, leveraging any standard team formation processes, to facilitate effective team operation. The objective is to ensure that all teams supporting a project receive the information, access, support and facilities necessary to achieve optimum performance.

"In the best work groups, the ones in which people have the most fun and perform at their upper limits, team interactions are everything. They are the reason that people stick it out, put their all into the work, overcome enormous obstacles."[30]

Tom DeMarco and Timothy Lister, Peopleware

The plan should address the means for addressing the factors that contribute to high performance and building team capability and confidence in those areas. The following fifteen factors are suggested as a starting point:

1. Clarity of team mission, vision, goals and priorities
2. Member and sponsor acceptance of team mission, vision, goals, priorities
3. Clarity of specific measures to track performance against goals
4. Amount of challenge in team's task
5. Team's value to members as a place to acquire new skills
6. Support for team from sponsor

7. Skill resources in team, including leadership
8. Team's size
9. Facilities, technology and support
10. Team's reporting relationship with sponsor
11. Ground rules on team operation, confidentiality, sign off, etc.
12. Team roles, boundaries and authority limits
13. Communication between this and other relevant individuals, groups
14. Team's measured performance against established goals
15. Stakeholder satisfaction.

Control Factor

The Control Factor within the Project Domain includes those control mechanisms that help stakeholders shape and guide the project to achieve the objectives of the planned change.

Over the course of a project, stakeholders will encounter many variances between the plan and actual results and will have decisions to make, including:

- o Stay the course.
- o Increase or decrease allocated funding or divert funding from one phase to another.
- o Increase or decrease staff complement or utilization.
- o Change key players.
- o Add, change or delete function, features and capabilities.
- o Modify quality targets
- o Change benefit forecasts or timing.
- o Revise phasing and staging strategies and plans
- o Celebrate completion.
- o Terminate the project.

Each decision taken on each Decision Area should have the support of all stakeholders and be recorded formally for communication to all interested parties.

The Control Factor includes the following Decision Areas: Release Plan Performance, Contracts Plan Performance, Risk Plan Performance, Team Plan Performance, Change Tracking and Reporting, Issue Tracking and Reporting, and Project Completion.

The Release Plan Performance Decision Area bundles the review of progress against a number of the Planning Factor Decision Areas together—Release Plans, Cost Estimates, Benefit Plan, Quality Plan and Resources and Facilities Plan—because of the inter-relationships and inter-dependence of these Decision Areas. However, the Contracts Plan and Risk Plan play a support role to the core project efforts. Because they are enablers and progress reviews can be conducted independently, separate Control Factor Decision Areas are included.

With the exception of the Project Completion Decision Area and unlike most others, all of these Control Factor Decision Areas should be revisited frequently over the course of the project. This will confirm the stakeholders' level of comfort and agreement with project progress and record any follow-on actions. The frequency of execution will be established by the Communication Plan.

Release Plan Performance

The Release Plan Performance Decision Area is concerned with managing project progress relative to the plans addressed in the Planning Factor Decision Areas: the Release Plans, Cost Estimates, Benefit Plan, Quality Plan and the Resources and Facilities Plan. Stakeholders need to address the following dimensions:

- o Ensure that the content (scope, function, capability) in each release is consistent with the plan.
- o Ensure that the actual completion of deliverables is in step with the baseline schedule.
- o Confirm that the actual costs are in line with the baseline estimates.
- o Verify that forecast benefits are being realized and banked according to the benefit plan.
- o Ensure that the quality targets are being achieved as planned.
- o Confirm that the required resources have been allocated and are being utilized according to plan.

Two techniques for managing progress that have been used widely and successfully—earned value and profile control—are outlined below.

Decision Framework

Earned Value

Earned value is a management technique that relates resource planning to schedules and cost. All work is planned, budgeted, and scheduled in time-phased "planned value" increments constituting a cost and schedule measurement baseline.

Earned value requires each task to have both entry and exit criteria and a step to validate that these criteria have been met prior to the award of the credit. Earned value credit is binary with zero percent being given before task completion and a 100 percent when completion is validated. As complex as it sounds, it's actually a very effective method for assessing progress against plan.

Profile Control

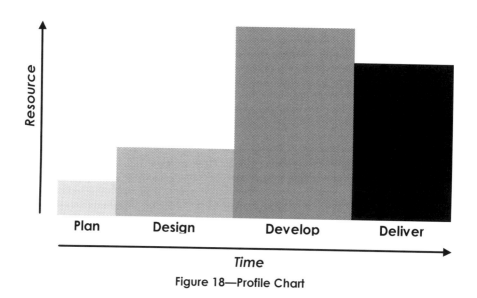

Figure 18—Profile Chart

A profile control is used to limit the time or expenditure on any given phase of a project based on organizational or industry experience with similar types of undertakings.

For example, the chart above shows a profile for systems development initiatives that allows approximately 10% of resource costs in the planning stage, 20% in the design phase, 40% in development and 30% in delivery (acceptance test, implementation, rollout and warranty effort). However, from a time, or schedule, standpoint, the profile allows 15% for planning, 30% for design, 30% for development and 25% for delivery.

Stakeholders can use the profile control to assess the reasonableness of project phase plans relative to the total estimated cost and to control the schedule and costs of each phase of a project based on an appropriate industry or organizational profile. It can provide an early warning where the estimated or projected time or cost for a given phase relative to the estimates for the total effort is out of line with the profile being used.

Here's an example of how valuable profile control can be in monitoring a project's progress.

Knowing Your Project Profile

The Situation

This Canadian insurance company was experiencing significant growth and having difficulty maintaining service levels at an affordable cost. The underwriting organization, in particular, was having a challenging time. Experienced underwriters were hard to find and training new staff was a long, slow process. The sales agents were unhappy with the response and level of quality on new applications for insurance. It cast the organization in a bad light with their clients. And, of course, agent compensation was negatively affected.

The VP responsible for the underwriting function, in consultation with the CIO, decided to explore the use of technology to alleviate the pressure, improve service and reduce costs.

The Goal

The VP of the underwriting function, the sponsor of the initiative, decided to launch a development project to automate much of the underwriting process. The plan was to have the applications for insurance processed by staff in the sales offices across the country using the new automated service with the potential for immediate approval and contract production if everything checked out. The target was to get the project implemented and operational in a year or less.

Decision Framework

The sponsor and his staff calculated that they could save over $1 million annually in reduced administrative costs in his organization while cutting processing time from 10 days to 1 day on 60% of the applications. As well, sales agent compensation would be accelerated, service to clients would be improved dramatically and dependence on scarce underwriting resources would be reduced.

In consultation with the Sales VP, it was determined that the additional time required from administrative staff in the sale offices to process the insurance applications could be factored into the existing workloads and would require no incremental staffing. The project was a no brainer!

The Project

The sponsor appointed a manager from his organization to act as overall project manager. His job was to implement new underwriting and administrative processes and practices that would affect the sales agents, administrative staff in the sales offices and underwriting and administrative staff at head office.

The CIO appointed a project manager from his organization to guide the development of the technology solution.

The two managers collaborated on the estimate based on their previous experiences and arrived at a cost of $2.5 million and a year or so duration. They agreed to adopt the in-house system development methodology, a typical waterfall practice. They formed a steering committee with the key stakeholders—the sponsor, the VP Sales and the CIO—and proceeded to form their teams to tackle requirements definition.

The sponsor was adamant that the new automated system reflect their current underwriting practices to maintain or improve on their claims experience. So, the project managers adopted a prototyping philosophy that they could us to demonstrate practice compliance to the sponsor and his senior underwriters as well as to the staff in the sales offices. As the requirements definition progressed, the project managers encountered a multitude of "what if" questions in response to their prototypes, mostly from the senior underwriters. That required another round of prototypes and reviews and lead to more "what if's", and so on.

The planned three month requirements phase extended to four months, then five but the project managers insisted the overall project was still on target. Their rationale was that the requirements would be so well defined and fully approved that the 10% change contingency they had included in the budget would not be required. They also argued that the level of detail in the prototypes would enable multiple parallel development streams and reduce the cost and time required for the development and testing phases.

The sponsor and the VP Sales bought the argument. The CIO did not. He had his Project Office pull together a profile for projects completed by the system development group in the previous two years. The profile showed that, on average, the development and delivery phases consumed over 60% of elapsed time and over 70% of resource costs. He concluded that the project would take another year to deliver and exceed the estimate by $1.5 million.

The Results

The CIO was right. The project took a year and a half to complete and cost almost $4 million. The IT project manager tried to reduce the elapsed time by developing the application components in parallel as he had argued previously. However, resource constraints limited the contribution the approach could make to accelerating the schedule.

The upside? The project was delivered successfully. The quality of the solution was excellent. The staff affected by the change in the sales offices and head office were enthusiastic about the system. The clients appreciated the service improvements. The project actually delivered over $2 million in annual benefits, largely because of the scope expansion resulting from the "what if" cycles. Even with the increased cost, the actual payback was greater than originally anticipated.

How a Great PM Could Have Helped

The project managers did a number of things well:

- Forming the steering committee helped all areas affected by the change get involved and onside.
- Use of prototyping to facilitate requirements definition really did engage the experts and resulted in full backing from all involved. It also did contribute to a much lower level of change requests and helped deliver significantly greater value than originally planned.

Decision Framework

- The quality practices leveraged from the in-house development methodology were applied effectively throughout the development process, ensuring a high quality solution in all respects, including the required ease of use for each target audience, appropriate performance and security, provision of an audit trail, integration with back end administrative systems, continuity of processing and adaptability to future changes.

Unfortunately the PM's missed a couple of opportunities to help the project excel and damaged their bosses' perceptions of their performance in the process:

- They didn't recognize that the "what if" cycles were signs of scope creep. There's nothing inherently wrong with expanding project scope but it has to be managed to expectations. In this case, it wasn't.
- There was a good deal of urgency in getting relief from the growing costs and service problems. But, the plan didn't recognize the need for a fast response. Instead, the PM's planned to define all requirements up front. Had they adopted a phasing strategy from the get go, they would have had an opportunity to deliver a first release sooner than planned and would have had a framework to manage the "what if" cycles more effectively.
- The IT PM started looking for resources to support his parallel development plan as the requirements stage wound down. Instead, had he really understood the urgency of the situation, he would have had a resource plan in place from the start with the required staff ready to step in when needed.
- Finally, the PM's didn't have a Project Profile for the organization in question. If they had access to that information, it would have forced more rigor on the cost and time forecasting and reduced the wishful thinking that they ultimately engaged in to justify their circumstances.

This could have been a great project! In fact, after the initial disgruntlement over cost overruns and schedule slippage, the stakeholders were extremely pleased with the results. The PM's could have been heroes. Instead, they were goats. It would have helped if the other stakeholders had challenged their approach and assumptions earlier. Unfortunately they didn't. The PM's missed the opportunity to take advantage of a few simple practices that could have made all the difference and they paid the price.

Risk Plan Performance

While the risk mitigation activities should be an integral part of the overall project plan and performance review, the risk plan itself should receive frequent attention from stakeholders. The focus should be on ensuring the following factors reflect the most current intelligence:

- o   All risks which could impact project performance have been identified.
- o   The owner of each risk has been identified.
- o   The potential impact on each release has been assessed and is current.
- o   The probability of the risk occurring has been assessed and is current.
- o   The trigger events that signal a risk becoming a reality have been identified and are still relevant.
- o   The priority of each risk has been established and is still appropriate.
- o   The mitigation plans and actions have been established for each risk and are still appropriate.
- o   Suitable contingency plans have been established and funding has been allocated where appropriate.
- o   Team Plan Performance

Stakeholders should be vigilant regarding the progress and performance of the teams created or being used to deliver the planned change.

Control should focus on the decisions made relative to the Team Formation Decision Area under the Resources Factor in the Assets Domain and the Team Formation Plan Decision Area under the Organization Factor on the Project Domain.

Change Tracking and Reporting

The Change Tracking and Reporting Decision Area requires stakeholders to apply the decisions made regarding the Change Control Decision Area to ongoing requests for change. The change requests can originate from a variety of sources, including:

- o   Enhancement requests from stakeholders
- o   Undiscovered task requirements
- o   Change agent and target initiated enhancements
- o   Market changes due to competitive, legislative, supplier, economic or other causes.

Decision Framework

Stakeholders should consider each change request in the context of the overall project and its impact on the other Decision Areas and decisions made. The final judgment, to approve, decline, amend or defer should be formally documented along with any ancillary impact on other areas.

Stakeholders should also look at the aggregate activity for change control requests, the sources of the requests, whether the requests are increasing or decreasing over time, the overall cost, schedule and benefit impact and the risk exposure from these trends.

Issue Tracking and Reporting

Issue management deals with any situation that may have an affect on the project's ability to deliver to project objectives, including any potential impact on benefits, costs, schedule or quality.

The Issue Tracking and Reporting Decision Area requires stakeholders to apply the decisions made regarding the Issue Management Decision Area under the Project Management Factor in the Assets Domain to issues raised.

As with Change Tracking and Reporting, stakeholders should also look at the aggregate activity for issue submissions, the sources of the issue, whether they are increasing or decreasing over time, the overall cost, schedule and benefit impact and the risk exposure from these trends.

Project Completion

When will your project be complete? What conditions need to be satisfied for completion to occur? Who will have to agree that the project is, in fact, complete and what expectations will they have? The last thing you want to happen is a debate on these questions as the project winds down. In fact, if you are debating these questions near what you think is the end of the project, chances are you've missed the target. Applying *FastPath* from inception through completion can avoid the debates and ensure your project wraps up with accolades and wins the gold medal.

Perhaps the ultimate project control step requires stakeholders to decide on the completion of a project. It is a decision! Completion doesn't happen just because the project actually implemented something, or because the sponsor says so, or because the staff have been committed to another project. Completion occurs when stakeholders agree that the goals established for the project have been realized or, that there is no hope of realizing those goals.

Ironically, the completion decision is not a point in time but a process that begins well before the project wraps up and progresses through to a final conclusion that can see:

o   Official termination all project activities
o   Reallocation of all resources—human, technology, physical plant, etc.
o   Finalization of all project dependent relationships—contracts, partners, suppliers, etc.
o   Tallying the final costs
o   Banking the benefits realized
o   Assessing and sharing the lessons learned
o   Initiating follow-on activity to close gaps or take advantage of new opportunities
o   Turning the delivered solution over to others to operate and support.
o   Monitoring performance, quality and stability of the delivered solution.
o   Managing warranty work to ensure the delivered solution satisfies service level agreements
o   Evaluating, recognizing and rewarding team and individual member performance.

The completion exercise can also relate to phase by phase implementations as well as premature terminations.

So how does the monitoring exercise work? It's pretty easy actually. The stakeholders start with a questionnaire which morphs into a scorecard reflecting the changes in their level of agreement over time. It's called the Project Pre-Check *FastPath* process!

Communication Factor

"Communication is the most important skill in life."
Stephen R. Covey—

Author of The Seven Habits of Highly Effective People

Effective communication is a critical success factor for optimum project performance. Yet we tend to think of the ubiquitous project status report whenever the topic of project communications is mentioned. In reality, it's probably the least important tool in the communication arsenal. Why? Do you know what information the other stakeholders need to feel comfortable and confident? Do they know what you need? Is there a common understanding of the individual needs of each stakeholder as it relates to content, timing, format, milieu, medium? Here are some suggestions on how to craft a communication machine to elevate project performance.

Monitor Effectiveness

The Communication Plan establishes the processes, activities, roles and responsibilities to identify, deliver and receive appropriate and timely communications. It also identifies all involved participants and organizations and requires participation by the staff in each organization to ensure a successful implementation.

The Monitor Effectiveness Decision Area requires that stakeholders assess the effectiveness of the formal communications based on the measures and frequencies established in the Communication Plan and take corrective action if the communications are not achieving the desired result.

Project communication needs to engender and leverage stakeholder relationships. Sure, all parties with a stake in a project require access to quality information, the right information, in the right format and form on a timely basis. They also need venues to vet issues and concerns and forums to share and reach consensus. Each stakeholder may also require their own unique views of content and progress to satisfy their personal communication styles and organizational responsibilities. But most importantly, successful projects need effective collaboration. To achieve that end, the right communication strategies and practices are essential.

There are as many project tracking and reporting practices as there are projects. Take a look at any standard practices within the organization and determine whether the existing standards can meet the needs of stakeholders and other constituents or at least serve as a foundation for specific additional requirements. To determine what your project needs, consider the following:

Monitor Updates

The Communication Plan establishes the processes, activities, roles and responsibilities to identify, deliver and receive appropriate and timely communications. It also identifies all involved participants and organizations and requires participation by the staff in each organization to ensure a successful implementation.

Great! So how is the plan working? Stakeholders need to assess the effectiveness of the formal communications based on the measures and frequencies established in the Communication Plan. If the plan isn't firing on all cylinders and the communications are not achieving the desired result, change the plan or fix the execution so that the gap is addressed.

Monitor Feedback

The Communication Plan will identify all the formal and informal means of providing information and establishing the necessary dialogues with project stakeholders. As described above, by monitoring the effectiveness of the plan, any issues with communication effectiveness can be addressed.

However, stakeholders should also receive direct and indirect feedback outside of the formal communication program. It will come from a variety of sources and relate to the progress of the change and how it is perceived by those directing, involved in and affected by the change. Monitoring feedback deals with these informal and unsolicited messages. It's about tapping into the informal "grape vine". It requires stakeholders to actively monitor and solicit feedback outside the formal channels, share the messages received with other stakeholders and take collective, corrective action as required.

Decision Framework

There you have it, the ingredients for superlative project communications and a critical catalyst for success; develop a communication plan that all stakeholders support, monitor the effectiveness of the program and revise as necessary and leverage insights gained from tapping into the grape vine.

"The most important thing in communication is to hear what isn't being said."

Peter Drucker

APPENDICES

A. Decision Area Sources
B. Additional Sources

A. Decision Area Sources

Project Pre-Check is about reuse—reuse of industry best practices, reuse of Project Pre-Check processes, reuse of the Decision Framework to filter best practices into Decision Areas and reuse of the Decision Areas to help stakeholders leverage the wisdom and insight of countless people and projects the world over to improve their ability to deliver change successfully.

The following organizations and practices have had a sizeable influence on Decision Area selection. Some are focused primarily on information technology; others target management of change, project management, software development, security and quality. It is this breadth and depth of diverse industry practice that gives the Project Pre-Check Decision Areas the brawn to support stakeholder decision making across an unlimited range of change possibilities.

COBIT

Control Objectives for Information and related Technology (COBIT) was created by the IT Governance Institute (ITGI). It provides a comprehensive framework for the management and delivery of high-quality information technology-based services.

CMMI

Capability Maturity Model Integration (CMMI) was created by the Software Engineering Institute at Carnegie Mellon University. It combines three source models—Capability Maturity Model for Software, Electronic Industries Alliance Interim Standard and Integrated Product Development Capability Maturity Model—into a single improvement framework for use by organizations pursuing enterprise wide process improvement.

DSDM

Dynamic Systems Development Method (DSDM) was created by the DSDM Consortium to provide a framework of controls and best practices for the rapid application development of high quality business system solutions. It does not say how things should be done in detail, but provides a skeleton process and product descriptions that are to be tailored to suit a particular project or a particular organization.

ISO/IEC

ISO/IEC is the International Organization for Standardization. The standards of primary concern to Project Pre-Check include ISO 9001:2000—Quality Management Systems and ISO/IEC 17799:2005—Code of Practice for Information Security Management

ITIL

The *IT Infrastructure Library* (ITIL) is a collection of best practices in IT service management from the Office of Government Commerce (OGC) in the United Kingdom. It is focused on the service processes of Information Technology management.

Management of Change (MOC)

A number of management of change practitioners influenced the Project Pre-Check Decision Areas including John Kotter, Conner Partners (formerly ODR) and Robert Schaffer. Their focus, like Project Pre-Check, is on delivering major business and technology change successfully.

MOF

Microsoft's *Microsoft Operation Framework* (MOF) provides technical guidance to organizations to achieve mission-critical system reliability, availability, supportability, and manageability of IT solutions built with Microsoft products and technologies. MOF's guidance addresses the people, process, technology, and management issues. MOF also combines the ITIL standards with specific guidelines for using Microsoft products and technologies and extends the ITIL code of practice to support distributed IT environments and industry trends.

MSF

Microsoft's *Microsoft Solution Framework* (MSF) describes a high-level sequence of activities for building and deploying IT solutions. It is a deliberate and disciplined approach to technology projects based on a defined set of principles, models, disciplines, concepts, guidelines, and proven practices.

OPM3

The *Organizational Project Management Maturity Model* (OPM3) from the Project Management Institute (PMI) offers an assessment process by which an organization can measure its organizational project management maturity and improve its current state.

Other Sources

A multitude of other sources have been used with positive results over the years and were referenced for best practice ideas in the development of the Decision Area catalogue. The sources include books, periodicals, business and technical magazines, web sites, newsletters, peers and personal experience. The Notes and References sections at the end of the book include many of the most worthwhile sources.

P3M3

The Portfolio, Programme & Project Management Maturity Model (P3M3) was developed by the UK Office of Government Commerce (OGC) to "help organizations address fundamental aspects of managing portfolios, programmes and projects, improve the likelihood of a quality result and successful outcome and reduce the likelihood of risks impacting projects adversely".

PMBOK

Guide to the Project Management Body of Knowledge (PMBOK), from the Project Management Institute (PMI) is described as 'the sum of knowledge within the profession of project management'. It is an American National Standard, ANSI/PMI 99-001-2004.

Prince2

Projects in Controlled Environments (PRINCE) provides a structured method for effective project management. It is produced and published by the UK Office of Government Commerce (OGC).

QAI

The Quality Assurance Institute (QAI) was founded in 1980. Its objective is to provide leadership in improving quality, productivity, and effective solutions for process management in the information services profession.

RUP

The *Rational Unified Process* (RUP), now owned by IBM, is an iterative software development methodology. It provides detailed descriptions of not just the 'what' needs to be done, but also 'how to' carry out its activities including guidelines and templates.

Val IT

Val IT, developed by the IT Governance Institute (ITGI), provides the means to measure, monitor and optimize the realization of business value from investment in IT. Val IT complements COBIT from a business and financial perspective and helps in value delivery from IT.

B. Additional Sources

The sources for the best practices that were assessed and used to drive out the Project Pre-Check Decision Areas are as varied as the collective experiences of the author and the relationships with hundreds of creative, talented IT and business professionals will allow. However, those sources are somewhat difficult for the reader to tap. So, listed below are a number of the more accessible sources for your reference.

Organization	Source
Atlantic Systems Guild	www.systemsguild.com/
Baseline Magazine	www.baselinemag.com/
CIO Insight Magazine	www.cioinsight.com/
Computerworld	www.computerworld.com/
Control Objectives for Information and Related Technology	www.isaca.org/cobit
Cutter Consortium	www.cutter.com/index.shtml
Dynamic Systems Development Method	www.dsdm.org/
Forrester Research	www.forrester.com/
Gantthead	www.gantthead.com/
Gartner	www.gartner.com/
Harvard Business Review	harvardbusinessonline.hbsp.harvard.edu
Institute of Electrical and Electronics Engineers, Inc.	http://www.ieee.org/portal/site/
Integrated Computer Engineering Inc	www.iceincusa.com/
International Function Point Users Group	www.ifpug.org/
International Standards Organization	www.iso.org/iso/en/ISOOnline. frontpage
IT Infrastructure Library	www.itil-itsm-world.com/
Ivey Business Journal	www.iveybusinessjournal.com
John Kotter	www.johnkotter.com/videos.html
Microsoft Operations Framework	www.microsoft.com/mof/

Appendices

Organization	Source
Microsoft Solution Framework	www.microsoft.com/msf/
Object Management Group	www.omg.org/
ODR Inc	www.connerpartners.com/
Office of Government Commerce	www.ogc.gov.uk/
Project Management Institute	www.pmi.org
Project Times Magazine	www.projecttimes.com/
Qualitative Software Management	www.qsm.com/index.html
Quality Assurance Institute	www.qaiworldwide.org/
Rational Unified Process	www-306.ibm.com/software/ rational/
Software Engineering Institute—	www.sei.cmu.edu/
Software Productivity Center	www.spc.ca/
Software Productivity Research	www.spr.com/
Software Program Managers Network	www.spmn.com
Spice	www.sqi.gu.edu.au/spice/
Standish Group	www.standishgroup.com/
Tech Republic	www.techrepublic.com

Clarke's Law of Revolutionary Ideas: Every revolutionary idea—in Science, Politics, Art or Whatever—evokes three stages of reaction. They may be summed up by the three phrases:

1. "It is completely impossible—don't waste my time."
2. "It is possible, but it is not worth doing."
3. "I said it was a good idea all along."

About The Author

R. Andrew Davison is principal consultant in Davison Consulting, a process consulting firm which focuses on project, program and portfolio management, strategy formulation, process and practice development and the implementation of high performance teams. Previously, he was an information technology practitioner, manager and executive in the financial services industry.

He studied history at Carleton University in Ottawa, Ontario, holds a fellowship from the Life Office Management Association and has nurtured a passion for continuous improvement throughout his career.

When he's not helping organizations improve their practices and performance, he can be found walking or biking the roads and trails around Long Point, Balsam Lake, reading, cooking (in a pinch) and enjoying the sunsets and a glass of wine with his wife Louise and Bearded Collie Tillie (who doesn't drink wine but likes the sunshine).

Index

Index

Index

Notes

1   MIS Quarterly Executive, *IT Project Management: Infamous Failures, Classic Mistakes, and Best Practices*, University of Minnesota, June 2007

2   The Standish Group International, Inc. *The CHAOS Report (1994)* as well as *(1999) Unfinished Voyages, A Follow-Up to The Chaos Report* from The Standish Group Web site: www.standishgroup.com

3   As reported on page 2 of the report from *The Federal IT Project Manager Initiative*, http://www.ocio.usda.gov/p_mgnt/doc/CIO_Council_Guidance. ppt#425,1, Federal

4   As reported in a February 7, 2003 TechRepublic article at http://articles.techrepublic.com.com /5100-10878_11-5034439.html

5   Ibid

6   Edward Yourdon, *Death March: managing "Mission Impossible" Projects*, Prentice Hall, 1997

7   Jennifer Stapleton, *DSDM: Dynamic Systems Development Method*, DSDM Consortium, 1997

8   Project Management Institute, *A Guide to the Project Management Body of Knowledge Third Edition*, 2004, p. 3

9   Ontario Ministry of Government Services, *http://www.mgs.gov.on.ca/english/ministry/releases/nr072805.html*, Queen's Printers for Ontario, 2005, p. 4

10 Office of Government Commerce, *Portfolio, Programme & Project Management Maturity Model* (P3M3), 2006, p.4

11 The Standish Group International, Inc. *Extreme Chaos* from The Standish Group Web site: www.standishgroup.com, 2001, p. 4

12 Ibid, p. 4

13 PMI, op. cit., p. 21

14 Ibid, p. 24

15 Edward H. Baker, *Hearts & Minds*, CIO Insight Magazine, Oct. 15, 2004

16 John P. Kotter and Dan S. Cohen, *The Heart of Change*, Harvard Business School Press, 2002,

17 Ibid, p. 41

18 Nicholas G. Carr, www.roughtype.com

19 James M. Kouzes and Barry Z. Posner, *The Leadership Challenge*, Jossey-Bass Inc., 1987, p. 8

20 Stephen R. Covey, *The 8th Habit*, Free Press, 2004, p. 5

Notes

21 Robert H. Schaffer, *Demand Better Results—And Get Them*, Harvard Business Review, March 1991,

22 John R. Hauser and Gerald M. Katz, *Metrics: You Are What You Measure!*, Massachusetts Institute of Technology, 1998,

23 Ibid, p. 1

24 Ibid, p. 12

25 Paul Glen, *Project Managers: Stop "gathering" IT requirements*, http://articles. techrepublic. com.com/5102-10878-6112248.html, Sept 5, 2006

26 Nick Jenkins, *A Project Management Primer*, http://www.nickjenkins.net/prose/projectPrimer. pdf, 2006

27 International Organization for Standardization (ISO), *ISO/IEC 17799:2005 Information technology—Security techniques—Code of practice for information security management*, http://www. iso.org/iso/en/prods-services/popstds /informationsecurity.html

28 Office of Government Commerce (OGC), *ITIL Service Support* and *ITIL Service Delivery*, http://www.itil.co.uk/

29 Microsoft, *Microsoft Operations Framework*, http://www.microsoft.com/technet/ itsolutions/ cits/mo/mof/default.mspx

30 Tom DeMarco and Timothy Lister, *Peopleware*, Dorset House Publishing Co., 1987, p. 121